A Chosen Calling

Medicine, Science, and Religion in Historical Context
Ronald L. Numbers, Consulting Editor

A Chosen Calling

Jews in Science in the Twentieth Century

NOAH J. EFRON

Johns Hopkins University Press
Baltimore

Hebrew Union College Press
Cincinnati

Johns Hopkins University Press
2715 North Charles Street
Baltimore, Maryland 21218-4363
www.press.jhu.edu

Library of Congress Cataloging-in-Publication Data

Efron, Noah J., author.
  A chosen calling : Jews in science in the twentieth century / Noah J. Efron.
     pages cm. — (Medicine, science, and religion in historical context)
  Includes bibliographical references and index.
  ISBN-13: 978-1-4214-1381-5 (hardcover : alk. paper)
  ISBN-10: 1-4214-1381-7 (hardcover : alk. paper)
  ISBN-13: 978-1-4214-1382-2 (electronic)
  ISBN-10: 1-4214-1382-5 (electronic)
    1. Science—Social aspects.   2. Judaism and science.   3. Jews—United States—Intel-
lectual life—20th century.   4. Jews—Russia—Intellectual life—20th century.   5. Jews—
Palestine—Intellectual life—20th century.   6. Jewish scientists—Social aspects—United
States—History—20th century.   7. Jewish scientists—Social aspects—Russia—History—
20th century.   8. Jewish scientists—Social aspects—Israel—History—20th century.
9. United States—Civilization—Jewish influences.   10. Russia—Civilization—Jewish
influences.   11. Palestine—Civilization—Jewish influences.   I. Title.   II. Title: Jews in
science in the twentieth century.
  Q180.55.S62E37 2014
  500.892'4009041—dc23   2013036532

A catalog record for this book is available from the British Library.

*Special discounts are available for bulk purchases of this book. For more information,
please contact Special Sales at 410-516-6936 or specialsales@press.jhu.edu.*

For Rachel, who taught me most of what matters

CONTENTS

A Vanload of Rabbis in the Culture Wars of Kentucky

T HIS BOOK BEGAN WITH an invitation to deliver the 2007 Gustave A. and Mamie W. Efroymson Memorial Lectures at the Hebrew Union College–Jewish Institute of Religion in Cincinnati, Ohio. For anyone who cares, as I do, about the history of Jewish life in America, HUC is a storied, almost mythological place. Begun by brilliant men whose mild manner and scrupulous scholarship masked the radicality of their ambitions (in particular, Rabbi Isaac Mayer Wise, who turns up in chapter 1 of this book), HUC has graduated generation after generation of leaders of the American Jewish community. Its library alone—with a Judaica collection surpassed only by that of Israel's National Library—is a site to which I had longed for years to make a pilgrimage.

I was not disappointed. The Teutonic erudition of HUC's founders finds lavish expression among the professors that fill the founders' endowed chairs and austere offices, putting me in mind of the first passage of *The Ethics of the Fathers*: "Moses received the Torah from Sinai and transmitted it to Joshua. Joshua transmitted it to the Elders, the Elders to the Prophets, and the Prophets transmitted it to the Men of the Great Assembly." I spent a quiet day viewing manuscripts in the archives. And, best of all, a van was procured and a trip organized to the Creation Museum in nearby Petersburg, Kentucky, for me, Hebrew literature professor and rabbi Susan Einbinder, and seven rabbinical students.

The museum had opened only months before, stirring controversy that I had followed with fascination in Tel Aviv. From the start, the museum struck me as an odd and subversive thing, adopting the syntax of a natural history museum to undermine the very notion of a natural history museum.[1] It was hard, at first, for me to understand what its founders and patrons had hoped to accomplish, and when I learned I would be nearby, I resolved to try to puzzle it out from up close. Doing that with a vanload of bright young Jews about to take their places at pulpits and in schools as leaders in Jewish communities around the country was an unexpected bonus.

Upon paying admission, we received with our change a brochure introducing the museum: "The purpose of the museum is three-fold," it read:

> First, it acts as a rallying place, calling people back to the absolute truth of the Bible. It is a place of revival, a starting point for a new reformation. Second, it is a witnessing tool. There will be those who sneer, but some will be challenged to think, and still others may come to believe. And finally, it is a valuable, unprecedented resource for information and education, enabling us to "always be ready to give an answer (a reasoned, logical defense) for the hope that is within us." (1 Peter 3:15)

I knew that when it opened, the museum had been received by some as a work of arrogance and nerve. More than eight hundred scientists from Kentucky, Ohio, and Indiana had signed a petition condemning it for ruining with knowing "misrepresentations" the chances of local kids to "succeed in science courses at the college level."[2] But as I began to make my way through the exhibits, it occurred to me that thoughtless arrogance is hardly the mood of the museum. The brochure described "calling people back" for "reformation" despite "sneers": this is not the program of confident crusaders. The Creation Museum did display bravado, braggadocio, and belligerence, presenting scientists as somnambulists, sheep, and charlatans—intellectual three-card-Monte grifters. But, then again, bravado, braggadocio, and belligerence—the marks of disgruntled adolescents—reflect insecurity, not confident superiority. As we ambled through its halls, the Creation Museum struck me less as a fanfare of triumphalism than a whimper of insulted indignation.

This was especially so in the museum's most disturbing exhibit, the "Culture in Crisis" display. Entering the exhibit, we were met with a wall of dense text and statistics about divorce and the decay of the family. Led through a simulacrum of a trash-strewn city street, in a piped-in soundscape of sirens and breaking glass, we were drawn to windows (in fact, video screens) through which we could peer voyeuristically into the daily doings of modern families. The scenes were sad. One showed two kids surfing porn while doing drugs. Another showed a teary teen clutching a pamphlet on abortion and dialing Planned Parenthood. A third showed a drunk man in a grimy undershirt berating his haggard wife. Across the way, a wrecking ball labeled "millions of years" had smashed the foundations of a church. I recognized the exhibit for what it was: an artless illustration of a view widely shared by evangelicals embittered by science, that godless theories of the world and its origins have swept the decency from our streets and homes.

As one preacher has put it, "Wherever evolution is believed, wherever it is taught, there it breeds chaos, revolution and immorality."[3]

Ken Ham, the charismatic founder of the Creation museum, edited a 2005 book called *War of the Worldviews* that on its back cover asks the question: "What do aliens, dinosaurs and gay marriage have in common?" It gives this answer: "They are all part of the culture war—a war between two worldviews. One view is based on a biblical understanding of history, the other on pure naturalism. Our educational institutions and the media are on the frontlines of evolutionizing our culture. From Biology 101 to World History, from The Learning Channel to SpongeBob, subtle and not-so-subtle evolutionary messages bombard us."[4]

In the end, what left the greatest impression on me from my visit to the Creation Museum was not the museum itself, but the reactions it provoked in my fellow visitors, those seven rabbinical students. At first, they surveyed the vitrines, LED screens, and elaborate dioramas with bonhomie and good humor, here making a joke, there offering glosses by medieval Jewish commentators to undermine the generally literal interpretation of biblical verses that informed the exhibits. But over time, the tenor of the chat grew serious. We came upon an odd exhibit: several columns of Torah parchment behind glass. The legend read:

Snatched from Saddam's Clutches

> Before he was removed from power in 2003, the Iraqi dictator Saddam Hussein ordered all Torah scrolls destroyed throughout Iraq. But several scrolls were smuggled out, such as this fragment, stuffed into the tire of a truck. Efforts to destroy God's Word have met similar resistance throughout history.[5]

I went white, and so did Susan and the students. One remarked that the inclusion of the leaves of Torah showed that the curators saw a link between evolution and anti-Semitism, and that Saddam Hussein's hatred of Jews had been co-opted into a battle in which it had no place. Jews had been co-opted into a museum in which we had no place.

After this, things that had earlier amused us—example after example of biblical verses interpreted literally and treated as scientific evidence on a par with laboratory data, or a putatively full-scale model of Noah's ark—became sources of unease. Over the hours we spent in the museum, our distress turned to despondency, and by the time we loaded into the van to travel back to the college, our despondency had blossomed into a sort of aching fear. "Something bad will come of this," one of the students said. The museum was a battering ram slammed against the wall of separation between church and state, and who could say how steadfast this wall would stand? We agreed that the museum did not offer its visitors so much as a molecule of anti-Semitism; inasmuch as it portrayed Jews at all, it did so with sensitive respect. But what the museum *did* offer, one of the students said, felt like a *child* of anti-Semitism and perhaps, then again, a *parent* of anti-Semitism. Everyone in the van agreed that the museum's attack on science was both an attack on the America in which we had all been raised, and that we loved, and also somehow an attack on the place of Jews in that America, or on the very notion that there *is* a place for Jews in that America. Although none of us could say precisely why, the Creation Museum's assault on science was somehow an assault on Jews, and this despite the fact that museum showed nothing but sincere and affectionate esteem for Jews.

This was an odd response, if you think about it, especially coming from a vanload of rabbinical students. These were women and men who had chosen a life of spiritual pursuits, holy texts, and communities of faith. Their calling is religion. One might expect from them, at the very least, some sympathy for the notion that religious explanations of the world deserve a place of respect in the public square. One might expect from them, at the very least, some appreciation of the angst behind the exhibits in the Creation Museum, and the fear that a world explained in secular and material terms alone is a world hostile to a life of righteousness. If a "culture war" is under way, as Ken Ham says, then one might expect rabbinical students to feel more *sympatico* for those on the side of religion than for, say, the author of *The God Delusion*, biologist and secular standard-bearer Richard Dawkins.

But as we drove back to Cincinnati, it became clear that if lines were drawn, the rabbinical students were on the side of the scientists and not the theologians. They (and I, if I am to be honest) were unnerved by the Creation Museum's agenda of bringing the Bible back into American schools. Their reactions to the museum, and mine, were no less unyielding than those of the scientists petitioning to close the museum and working nights to keep religion out of the public square. They, like me, saw in science something crucial, and perhaps fragile, that needed to be protected; something essential to our lives as citizens in a modern democracy, but somehow also something essential to our ability to live as *Jews*.

This attitude—our finding in science something of crucial value to us as

Jews—is one with a history that reaches back at least several generations, to the time when both modern science, as we know it, and modern Jewish life in the West, as we know it, took form. The pages that follow represent an effort to understand why this attitude took root when it did and where it did.

This book aims to offer not a survey (of a subject too vast to capture exhaustively) but an interpretation. It was born of a series of lectures, and though most of what follows was never uttered from a lectern, some of the ease that lectures offer survives in the text. The lectureship provided a rare opportunity to ask and try to answer a big question: why have the histories of modern Western science and modern Western Jewry been tightly bound since the start of the twentieth century? This question, in one form or another, has been a focus of decades not just of scholarly study but also of idle musing and countless conversations in seminar rooms, in synagogue sanctuaries, and around Shabbat tables. As in all attempts to puzzle out a big question, in tackling this one we risk overreaching. But there are some things that can be understood only if one is willing to risk overreaching. If in response to this book others offer alternative interpretations, additional facts, and different perspectives, leaving my own partial or obsolete, that will be the greatest compensation I could receive.

As I have said, my time at Hebrew Union College was a delight, owing to the delightful people I found there. If there are greater pleasures than hearing Susan Einbinder read the medieval poetry she has just discovered, they escape me. Years earlier I had learned that she is brilliant and enchanting, but it was in Cincinnati that I learned she is also a brilliant and enchanting host. David Gilner was generous with his time, wisdom, and enthusiasm for the remarkable collection he manages and continues to cultivate. Barry Kogan offered intellectual stimulation and inspiration. The Very Rev. James A. Diamond, may his memory be blessed, hosted one of my lectures in Christ Church, of which he was dean, and offered the best of what one hopes for in ecumenical interaction: an open heart and mind, and great kindness.

Of course, my intellectual debts range far beyond the city limits of Cincinnati. Ideas and insights in this book took form in conversation, over years, with a great number of remarkable scholars: Denis Alexander, Ian Barbour, Ari Barell, Sarit Ben-Simhon, Jeremy Benstein, Marshall Brinn, Jed Buckwald, Philip Clayton, Ron Cole-Turner, Pranab Das, Nadav Davidovich, Thomas Dixon, Willem B. Drees, Lia Ettinger, Raphael Falk, Carl Feit, Maurice Finocchiaro, Owen Gingerich, Tal Golan, William Grassie, Peter Harrison, Brian Heap, David Hollinger, Wentzel Van Huyssteen, Jonathan Israel, Eva Jablonka, Shaul Katz, Christoper Knight, Joe Kruger, Josh Lack,

Edward Larson, Hillel Levine, Bernie Lightman, David Lindberg, David Livingstone, Nancey Murphy, Margaret Osler, Channa Pinchasi, Bob Pollack, Larry Principe, Jon Roberts, Michael Ruse, Avi Sagi, Norbert Samuelson, Dov Schwartz, Eilon Schwartz, Sam Schweber, Michael Shank, Evelyn Simha, Ofer Sitbon, Bill Slott, Matthew Stanley, George Stocking, Marc Swetlitz, Alon Tal, Hava Tirosh-Samuelson, Jim Voelkel, Michael Walzer, Fraser Watts, and David Wilson. Three scholars in particular have been my guides and inspiration, my trinity, in the matters this book addresses and more: John Hedley Brooke, Geoffrey Cantor, and Ron Numbers. My debt to them is incalculably large. And two others have been my mentors, setting before me when I was young the questions that would absorb me throughout my life: Menachem Fisch and Rich Schuldenfrei. Day to day, hour to hour, I am richer for their generosity, creativity, decency, and brilliance. Looking at this list is a humbling affair; I have matched the insight and creativity of none of its members, but at least I have the good sense to know how lucky I have been to learn from them.

My good fortune extends too to my remarkable colleagues in the Graduate Program in Science, Technology & Society at Bar Ilan University: Raz Chen-Morris, Oren Harman, Nurit Kirsh, Boaz Miller, Anat Leibler, and Boaz Tamir. Working with them is a joy and frequently a revelation. It extends as well to our many gifted students, of whom several have contributed directly to this book: Gaby Admon-Rick, Dan Bar-El, Anat Ben-David, Erela Ben-Shahar, Shmuel Chayen, Moshe Cohen, Yuval Dror, Zvia Elgali, Ido Hartogsohn, Eyal Katvan, Ami Maoz, Avraham Menkes, Eynat Meytal, Naftali Moses, Eitan Nicotra, Vered Oppenheimer, Adi Paz, Rachel Pear, Dan Perry, Hemy Ramiel, Raz Schwartz, Hananel Serri, Tamar Sharon, Ilya Stambler, Dov Terkieltaub, Uri Weinhaber, Oshrit Yikneh, Adi Zamir-Nitzan, and Ariel Zimerman.

I am especially grateful for the care, dedication, expertise, and kindness exhibited to me by two publishing houses: Hebrew Union College Press and Johns Hopkins University Press. This book would not exist were it not for the encouragement and scholarly counsel of Michael Meyer, who until recently served as chairman of the Hebrew Union College publication committee. David Aaron and Angela Roskop Erisman, the new director of scholarly publications and the managing director of the press, respectively, displayed similar generosity upon Professor Meyer's retirement. Throughout the years of work on this book, managing editor Sonja Rethy provided help and inspiration, as well as countless suggestions and emendations that improved what the book says and how it says it.

It was a shot of tremendous good luck that I was able to work with

Jacqueline Wehmueller, an executive editor of Johns Hopkins University Press. Within her perfect professionalism is gentility and grace that meant a great deal to me. It was she who gave the manuscript to Carolyn Moser, who edited it with wisdom and care. It is a better book for her efforts. Thanks, too, to Andre Barnett, who brought the book to print with kind patience, and to Bob Land, fellow traveler in matters Reed and Cale, for preparing the index.

Early versions of portions of this book were presented in lectures at the Institute for Advanced Study, Princeton University; the University of Pennsylvania; Iowa State University; and the University of Leeds. A version of the chapter on science and technology in Palestine and Israel was published in *Zygon*, and an early version of a small part of chapter 1 was published in the *Reilly Center Reports*. Other portions are greatly expanded from my earlier book *Judaism and Science: A Historical Introduction*.

At the end of the day, what matters is the end of the day, which I am blessed to spend with Susan Warchaizer, Dara Warchaizer-Efron, and Micha Efron-Warchaizer. Tolstoy was wrong: their company is like no other, a puzzle, a joy, and a miracle. I was blessed, too, in my parents, Rosalyn and Herman Efron, my guides in so many things; as well as my brother, Adam Efron; and my sister, Rachel Efron. It was Rachel who flew out to Cincinnati to provide company and inspiration during my first foray into the American Midwest, only one installment in a lifetime of company, love, help, instruction, conversation, joy, insight, reflection, wisdom, and inspiration. It is to her that this book is dedicated.

A Chosen Calling

"Ridiculously Disproportionate"?

I N 1919, WHEN he turned his attention to the achievements of Jews in modern science, the great economist and social theorist Thorstein Veblen was sixty-two years old, weary, and despondent. Twenty years had passed since he had published the book that made him famous, his notorious attack on America's well-heeled, *The Theory of the Leisure Class*. Since then, he had left his first wife in an asylum, never to see her again, and had remarried only to witness his second wife's breakdown and internment in McLean Hospital in Belmont, Massachusetts, where, in 1920, she died as he stood at her bedside.[1] He quit university life in a huff twice, leaving Stanford in 1909 and the University of Missouri in 1918, the second time taking a job as the editor of *The Dial*, a New York politics and arts magazine with a socialist bent.[2] These were productive years for Veblen (remarkably so given the tumult of his personal life and that of the Great War, the Russian Revolution, and more), but what he produced was pessimistic. In 1918, he published *The Higher Learning in America*, a zealous condemnation of the destructive effect of businessmen-donors on weak-willed university administrators reluctant "to forego their habitual preoccupation with petty intrigue and bombastic publicity."[3] He was also at this time hard at work on *The Engineers and the Price System*, arguing that society will improve only when scientists and engineers join forces with working men against the monied and self-serving "Guardians of the Vested Interests."[4] In 1918, he published an essay in *The Dial* arguing that nation-states themselves "serve substantially no other purpose than obstruction, retardation and a lessened efficiency." The effort of workers to dismantle the nation-state, he continued, is at the heart of "that prospective contest between the vested interests and the common man out of which the New Order is to emerge, in

I

case the outcome of the struggle turns in favor of the common man."[5] This statement holds a key to much of the rest of his writing, and to Veblen's ragged radicalism. The Great War, the Bolshevik Revolution (which Veblen admired),[6] the rise of Zionism (which Veblen admired and regretted)[7]—all these were harbingers of a New Order that might emerge, were it not for "vested interests" and entrenched elites who were preventing it. It was the end of an era, Veblen saw, or at least it should and could be.

It was with the weight of all this on his heart that Veblen wrote "The Intellectual Pre-eminence of Jews in Modern Europe." His premise, he was sure, was undeniable:

> It is plain that the civilization of Christendom continues today to draw heavily on the Jews for men devoted to science and scholarly pursuits. It is not only that men of Jewish extraction continue to supply more than a proportionate quota to the rank and file engaged in scientific and scholarly work, but a disproportionate number of the men to whom modern science and scholarship look for guidance and leadership are of the same derivation. Particularly is this true of the modern sciences, and it applies perhaps especially in the field of scientific theory, even beyond the extent of its application in the domain of workday detail. So much is notorious.[8]

Veblen wondered why this was. The explanation he arrived at was that Jews had only recently freed themselves from the oppressive influence of their rabbis and traditions, and had not yet been sufficiently accepted in Christian society to come under the thumb of its ministers, politicians, and fat-cats—the same "vested interests" that were destroying the universities, bedeviling scientists and engineers, and suppressing the working stiff. Jews were the free radicals of Western civilization. Unbound, they developed an attitude of skepticism toward received pieties. They excelled at irony and alienation. These traits, Veblen argued, were precisely the traits required to excel at science: "The first requisite for constructive work in modern science . . . is a skeptical frame of mind." Unmoored, "a wanderer in the intellectual no-man's land," the Jew "becomes a disturber of the intellectual peace," which is just what it takes to overturn the laws of physics or rewrite the book of life or reinvent the human psyche.[9]

Veblen may have been the first scholar to ask why Jews succeeded extravagantly in modern science, but he was not the last. His explanation was controversial when he first offered it and remains so to this day.[10] The same can be said of the many other explanations that have been floated in the years since Veblen's essay appeared. In 1969, famed British novelist and physicist Charles Percy Snow hypothesized that Jews had perhaps evolved a *genetic* predisposition to succeed in science. In a speech delivered at Hebrew Union College in New York, "judging Jewish achievements on the basis of Western scientific culture," he found that "the Jewish performance has been not only disproportionate, but almost ridiculously disproportionate." All this left him "prepared to believe" that there is "something in the Jewish gene pool which produces talent on quite a different scale from, say, the Anglo-Saxon gene pool."[11]

Snow speculated that Jews might be better at science because they are *smarter* at science and that they might be smarter because of their genes.[12] He left it to others to explain how Jews got these superior genes, and they did. Norbert Wiener, a Jewish mathematician who won a Bocher Prize and a National Medal of Science, and who had reason to believe that he was a descendant of Maimonides,[13] offered this explanation:

> The young [Jewish] learned man, and especially the rabbi, whether or not he had an ounce of practical judgment and was able to make a good career for himself in life, was always a match for the daughter of the rich merchant. Biologically this led to a situation in sharp contrast to that of the Christians of earlier times. The Western Christian learned man was absorbed in the church, and whether he had children or not, he was certainly not supposed to have them, and actually tended to be less fertile than the community around him. On the other hand, the Jewish scholar was very often in a position to have a large family. Thus the biological habits of the Christians tended to breed out of the race whatever heredi- tary qualities make for learning, whereas the biological habits of the Jew tended to breed these qualities in.[14]

Such reasoning amounted to a "sociobiological theory of Jewish intellec- tual achievement,"[15] a view captured well in 1966 by economist and eugeni- cist Nathaniel Weyl, who wrote that "Jewish intellectual eminence can be regarded as the end-result of seventeen centuries of breeding for scholars."[16] In 2005, researchers Gregory Cochran, Jason Hardy, and Henry Harpend- ing made the front page of the *New York Times* when they published a paper called "A Natural History of Ashkenazi Intelligence," in which they theorized that high-IQ jobs like money-lending and rabbinics left the smart- est Jews fittest to survive. Natural selection and sexual selection, working tirelessly in a unique cultural milieu that preferred pasty over ruddy, quick- witted over fleet-of-foot, and *Sitzfleisch* over *Schönheit*, had produced a people well-suited for laboratories and chess.[17] Arguments of this sort are buoyed by an assumption that high native intelligence is, by itself, enough to explain hearty and sustained success in science. This assumption may be true (though I have my doubts); still, one might wish to see it demonstrated.[18]

Of course, one does not need to believe that the value Jews have long placed on learning (ostensibly, at least) changed Jewish genes to think that it somehow gave Jews an edge in the laboratory. Another explanation of Jewish achievements in science has been that an enduring tradition of learning, and a love of learning, led Jews in droves to seek the advanced university degrees that lay the groundwork for scientific genius. Books, lectures, debates, and long, quiet hours of study have been a preferred pastime of Jews for centuries, this view goes. What modern Jews like to study may have evolved from Bible to biology and from commentaries to chemistry, but the *habits* of study that once produced towering Talmudic scholars can just as easily churn out scientific geniuses. After all, Jews are famously "the people of the book." To people who attribute the successes of Jews in modern sciences to their long-standing tradition of respecting study, it hardly matters whether the book in question records the laws of *halakhah* or the laws of thermodynamics.

Others hypothesized a deeper link between traditional learning and scientific excellence. Writer and Cambridge don George Steiner believed that modern Jews are heirs to a particular *style* of thought that developed in the yeshivas of past centuries. Generations of struggling to interpret canonical texts left Jews with a "commitment to analytic totality, to the ordering of all phenomena under the unifying laws and principles of prediction," that accounts in part for "the formidable contributions of Jewish thinkers, scholars and artists to the sum of the modern."[19] Others see the key to Jews' successes in science as arising from the Talmudic tradition of disputation. Classical Talmud study, to this day, is a meeting of two students who criticize, find fault, troubleshoot, and one-up each other's interpretations of the text before them. Unceasing efforts to *falsify* whatever your study partner puts before you, and to improve it, bear a resemblance to what some philosophers have described as the practice of science at its best.[20] Facility with Talmud study, some suggest, may give Jews a leg up in the laboratory, where questioning assumptions and criticizing accepted wisdom can sometimes produce surprising new theories. Gidi Grinstein, head of one of Israel's leading think tanks, recently told journalist Linda Gradstein that the key to Jews' success in science resides in a different aspect of their intellectual heritage: "For thousands of years, Jews have been brought up to question and to try to bridge the gap between existing knowledge and the prevailing reality. You have the Torah and the Talmud, and then you have the reality, which keeps changing. The tension between what we know and what we experience is the secret of creativity."[21]

There are problems with explanations of this sort, which attribute Jews' successes in science to their genes or their age-old traditions of learning. For one thing, they fail to account for the fact that, with a few notable exceptions, Jews showed little interest in science and no special talent for it until the end of the nineteenth century and, especially, the first decades of the twentieth. When the great Jewish folklorist Joseph Jacobs set out in 1886 to compare the talents of Jews with the talents of other Westerners, he did not find that they excelled in any science save medicine: "Jews have no distinction whatsoever as agriculturalists, engravers, sailors and sovereigns. They are less distinguished than Europeans generally as authors, divines, engineers, soldiers, statesmen, travelers. The two lists are approximately equal in antiquaries, architects, artists, lawyers, natural science, political economy, science, sculptors. Jews seem to have superiority as actors, chess-players, doctors, merchants (chiefly financiers), in metaphysics, music, poetry and philology. On the whole, these results correspond with the rough inductions of common experience."[22] It is easy to forget that when Veblen asked why Jews excelled at science, the phenomenon he hoped to explain was new and surprising. But if Jewish preeminence in the sciences was the result of their biology or their age-old heritage, why did these kick in only at about the start of the twentieth century?

Once this is recognized, theories that attribute Jewish scientific talent to this or that Jewish intellectual gift or tradition of study falter under their own weight. Upon reflection, it is not clear how (or, in the end, whether) the high regard Jews had for yeshiva study was transformed into equally high regard for university and laboratory study. It is not clear how or whether styles of Talmudic disputation were absorbed by secular Jewish scientists who themselves never cracked open a tractate of Talmud. And even if these styles were somehow absorbed, it is unclear how or whether they aided scientists in pursuing their research. Even if one were willing to brush aside the great difficulties that attend a genetic theory of Jewish scientific achievement —Can intelligence really be inherited? Could the homespun eugenics hypothesized by the theory have achieved a meaningful rise in Jewish intelligence during the course of the several dozen generations it was practiced? Are money-lenders really smarter than the poor saps forced to use their services? Does intelligence alone really explain success in science? and many more questions along these lines—it is not clear why Jewish success in science expressed itself so suddenly and why is it now declining, as recent studies have begun to show.

What's more, few of the Jews who made a name for themselves in science had ever learned much about their religion, and fewer still had studied sacred Jewish texts. The vast majority of Jewish scientists—in America, in Russia or the USSR, in Palestine or Israel, or wherever else they found themselves—were what writer Isaac Deutscher memorably called "non-Jewish Jews." Of "Spinoza, Heine, Marx, Rosa Luxemburg, Trotsky and Freud," and the multitude of Jewish intellectuals they represent, Deutscher wrote: "They all went beyond the boundaries of Jewry. They all found Jewry too narrow, too archaic, and too constricting. They all looked for ideals and fulfillment beyond it, and they represent the sum and substance of much that is greatest in modern thought, the sum and substance of the most profound upheavals that have taken place in philosophy, sociology, economics, and politics in the last three centuries."[23] It was from this intellectual tradition that most Jews who made their name in science came.

Almost a century has passed since Veblen asked what accounts for the remarkable preeminence of Jews in Western science and scholarship and offered his answer. During this time, Jews' preeminence grew to a degree that Veblen himself could never have predicted. As I have described, alternative explanations of the phenomenon have been offered since Veblen, and while there is likely some truth to several of them, just as there is some truth to Veblen's theory, the success of Jews in modern science remains almost as mysterious today as it was a century ago, when it first became apparent.

There are reasons that this is so. Berkeley historian David Hollinger—who has written brilliantly about the changing relationships between church, university, laboratory, and the public square in the twentieth century, and is our best guide in these matters—observed that Veblen failed to see the historical specificity of the phenomenon he described, taking it "too much for granted that history would constantly repeat [and] that Jews were 'forever' destined to be pariahs when in gentile society."[24] The tendency to see Jewish success in sciences in ahistorical ways is, in fact, common to almost all who attempt to explain it (most of all, those who see this success encoded somehow in proteins on chromosomes). We are further hobbled, Hollinger writes, by "the booster-bigot trap, which quickly channels discussions of Jews in comparison to other groups into the booster's uncritical celebration of Jewish achievements or the bigot's malevolent complaint about Jewish conspiracies."[25] It is difficult, nearly a century of flawed hypothesizing has shown us, to account for the success of Jews in the sciences without valorizing or demonizing.

To all this, I would add something else. As we have seen, many scholars (and a great many more of us, too: rabbis, Hebrew school teachers, bloggers, idle friends chatting over coffee or a Shabbat dinner) have followed Veblen in asking why Jews have been so remarkably successful in modern sciences. It may be, however, that this question cannot be answered until prior questions are posed: Why have the sciences been so remarkably successful among modern Jews? Why did so many twentieth-century Jews in so many places seek in the sciences a career and a vocation? Why did American rabbis lavish praise upon scientists from their pulpits? Why did American Jewish philanthropic foundations so commonly support research hospitals and research universities, science museums and planetariums, and other institutions that advanced science or the public appreciation of it? Why did the children of petty traders in the Pale of Settlement imagine their futures in Soviet laboratories, hospitals, and research centers? Why did Zionist pioneers envision a future in which scientific agriculture, industry, and medicine would transform the ageless Levant into a New Atlantis? What accounts for the great purchase science found in the imaginations of so many Jews, as well as in what they believed, how they lived their lives, and what sort of world they hoped to leave their children?

These questions do not have a single, uncomplicated answer. For one thing, as Hollinger insists, their answers differ from time to time and place to place. The twentieth century was, for Jews, a century in motion—a century of migration, emigration, and immigration. Historian Yuri Slezkine described these great movements schematically: "In the early twentieth century, Jews had three options—and three destinations—that represented alternative ways of being modern: one that was relatively familiar but rapidly expanding and two that were brand-new."[26] The familiar option was the liberal capitalism of America. The two new options were Palestine and the cities of the new Soviet Union. ("Most of the Jews who stayed in revolutionary Russia," writes Slezkine, "did not stay at home: they moved to Kiev, Kharkov, Leningrad and Moscow, and they moved up the Soviet social ladder once they got there. Jews by birth and perhaps by upbringing, they were Russian by cultural affiliation and—many of them—Soviet by ideological commitment."[27]) Like Slezkine, my attention, in the pages that follow, is focused on these three great destinations, at the time when new Jewish communities were quickly, and with surging energy, taking form in each. The full picture was more complicated still, and, as Slezkine readily acknowledges and I must as well, this focus leaves out many small Jewish communities spread throughout the West, and indeed across the globe. Each adoptive home brought its own challenges, and each Jewish community developed in its own way. The circumstances that brought a teenager from Cleveland to Cornell differed from those that brought a student from Kiev to Kharkiv University, or from Tel Aviv to the Technion. Different times and different places produced different stories. This is no surprise.

What is a surprise, though, is how much these different stories have in common. Jews were overrepresented in the sciences to similarly great degrees in the United States and in the Soviet Union (although, for obvious reasons, this sort of demographic measure has little meaning in regard to Jewish Palestine). In all three places, one finds similarly enthusiastic rhetoric about the role of science in fashioning a good society. In all three places, the "values" of science were said to comport with and bolster the most dearly held social values: democracy, equality, progress, and more. In all three places, Jews held a generally uncritical attitude toward the sciences and technology, and toward government by scientific principles. In all three places, a great many Jews displayed a great deal of enthusiasm for science, in its many forms and applications.

This commonality of experience among Jews in such varied circumstances is what spurs some to explain Jews' affinity for sciences as a product of their genes, culture, or history. If Jews in such divergent situations all gravitate toward science, this thinking goes, it must be a result of the very nature of the Jews themselves. This reasoning is understandable enough, but it is mistaken. Instead, the common experience of Jews in Boston, Birobidzhan, and Beer Sheva was a result of commonalities and similarities in the experiences even of many Jews who in so many ways differed.

In the late nineteenth century and the early twentieth, when considering the place of Jews in Western society, scholars, critics, pundits, and politicians (Jews and Christians alike) frequently described their efforts as attempts to solve what they called "the Jewish Problem" or answer what they called "the Jewish Question."[28] Although by the middle of the twentieth century these formulae had fallen into disrepute and disuse,[29] in earlier decades almost no one doubted that throughout the West there was such a "problem" and "question." Countless people tried to describe it. Writing in the *Atlantic Monthly* in 1941, Libertarian scholar Albert Jay Nock offered perhaps the most generic definition of "the Jewish Problem," which he characterized as "that of maintaining a *modus vivendi*" between Jews and the non-Jewish majority among whom they lived.[30] This was a problem faced by Jews wherever they settled in the twentieth century. At different moments, in different places, differing solutions were found, tried, and at times adopted. But the problem of establishing this *modus vivendi* was shared by Jews everywhere.[31]

As the chapters that follow show, modern sciences seemed to many Jews in many different circumstances to be a part of a solution to this problem. To achieve a *modus vivendi* with the societies in which Jews found themselves, some judged it necessary to advance two reform projects of staggering ambition. The first was a reform of Jews themselves, from parochials to full participants in the broader cultures in which they found themselves. The second was the reform of these broader cultures, along lines that would enable Jews to participate in them. Science was a way to do both things at once. For those who pursued it as a career, it was as *unparochial* a pursuit as any modern Western society offered. For those Jews who simply applauded or supported science from the side, this too was a way to cast their lot in with universal *human* advancement.

No less important, science promised to refashion the societies in which Jews found themselves along a more accepting pattern. Vladimir Lenin, the architect of the Soviet revolution and founding father of the Soviet Union, insisted that anti-Semitism was inconsistent with the scientific principles of government upon which the new nation was based. Science itself demanded nothing less than absolute equality among citizens. In America, many were coming to similar conclusions. The Conference on the Scientific Spirit and the Democratic Faith that met for several years at the Jewish Theological Seminary and Columbia University in the early 1940s set for itself the goals of protesting against "a movement of reaction" in America that was "inimical to both democracy and science," and advancing both the scientific spirit and democratic faith. It was around this time that Jewish philosopher Ernest Nagel attacked in the *Partisan Review* those "malicious philosophies of science" that ignored the tight intertwine of values between science and democracy. He and many others were certain that neither could thrive without the other.[32] In Jewish Palestine, the issues were different. Still, even there, Zionist ideologues and settlers looked to the sciences to fashion out of the rough-hewn wilderness of the Levant a progressive, pluralist society. Already in his early utopian novel, *Altneuland*, Theodor Herzl emphasized that the technological and scientific preeminence of the society he foresaw, and the strength of its democracy and pluralist toleration, were inseparably linked.

It ought not surprise us that in all of these very different circumstances, Jews looked to science to help refashion the societies in which they lived; as sociologist Yaron Ezrahi has observed, at this time people of many differ-

ent persuasions pervasively looked "to science to depoliticize or depersonalize political decisions and actions."[33] If you are a member of an ethnic or
religious minority seeking a voice in a society from which you were previously excluded, *depoliticizing* and *depersonalizing* political decisions and
actions is precisely what you want.

Which is not to say that there were not, during the first half of the
twentieth century, important differences in the value that Jews in different places accorded sciences and in how they hoped that sciences might
alter the societies in which they lived. The following chapters show that
there were great differences that mattered a lot. Still, Jews throughout the
West at this time faced the challenge (and the "problem") of creating a
new *modus vivendi* in the societies in which they found themselves. The
solutions they reached varied from place to place. But in every case, science was an invaluable tool. And this was because science as it developed in
the first half of the twentieth century was beautifully suited to the particular
needs of large numbers of Western Jews. Science was a national endeavor—a
pathbreaking scientist was a ready source of patriotic pride—that rewarded
a transnational perspective. It was at once parochial and universalist. Jews
in science served homeland by serving humanity—homeland and humanity
being the two great abstractions in which Jews had until not long before
been denied participation around the globe.

What's more, science promised not just an avenue for Jews into the societies in which they found themselves; it promised, too, to transform these
societies from the inside, leaving them more porous and permeable, less
able to justify keeping Jews apart. The values that early twentieth-century science increasingly championed—objectivity and meritocracy foremost among
them—fit perfectly with the needs of a minority people regarded with suspicion and anxious to make their way.

And if all this is true, then Veblen's thesis about Jewish preeminence in
science can be turned on its head. Rather than reflecting Jewish virtuosity
in irony and aptitude arising from alienation from the societies in which
they found themselves, the powerful appeal that sciences held for many
Jews may have resulted, in part, from the wish of these Jews to assimilate
into those same societies. Veblen attributed Jewish creativity in science to
the apathetic disregard of Jews for hidebound civility and the conformism
it fostered, and their easy willingness to thumb their noses at a society they
had little wish to join. But a more potent explanation of the attractions science held for many twentieth-century Jews—in America, Europe, and the
Levant—may be precisely the opposite: it helped them to *join* the societies in which they found themselves. If Veblen's Jews were connoisseurs of

alienation, those portrayed in these pages wished for nothing so much as to escape alienation, to fit in. To do this, they needed to amend themselves and amend the places they lived. And for a variety of complicated reasons, science promised to help them do both.

It will already be clear, I suppose, that in this book I make no attempt to describe all those Jews who made a mark (even a very significant one) in this or that science. Because of the prodigious success of Jews in twentieth-century science, doing so is very nearly impossible. What's more, it would be tiresome; save as an inducement to empty, we-rock! ethnic pride, it is hard to see its value. After years of study, I have found nothing *Jewish* about the science of Jewish scientists; their work itself says nothing that I can divine about their faith or heritage. But if the chapters that follow do not systematically portray Jews in science, they aim to portray the growing importance of science for Jews, and not just the scientists among them. They aim to capture the *enthusiasm* for science, the belief in science, the devotion to science that all sorts of Jews demonstrated, from the loftiest heights occupied by Einstein to the lowland of the Orthodox boy I was, reading *The Microbe Hunters* by flashlight under the covers after bedtime. This enthusiasm—which of course must be part of any account of why so many Jews chose science as a vocation and devoted themselves to it with single-minded passion—has much to tell us about the circumstances in which these Jews found themselves during much of the past century, and especially during the first half of the twentieth century, which is the major focus of this book.

I was young, only seven or eight by my estimate, when my mother told me the following joke: "What do you call a Jew with a master's degree?" Answer: "A dropout." At the time, I did not know what a master's degree was, and I did not know what a dropout was. But I was beginning to understand what a Jew—in America, in the middle of the twentieth century—was. One way to see the chapters that follow is as a midrash on my mother's joke.

CHAPTER ONE

"Holding High the Torch of Civilization"

American Jews and Twentieth-Century Science

IN THE SUMMER OF 1925, America was riveted and riven by the trial of John Thomas Scopes, a Dayton, Tennessee, high school teacher accused of teaching evolution in violation of state statute. Although many saw the trial as a skirmish between science and religion, less than two months after a guilty verdict was handed down the leading rabbis of New York City joined forces to condemn Scopes' conviction.[1] There was drama in the unanimity of the condemnation, and it was amplified by the rabbis' choice of venues: the pulpits of congregations throughout the town, on the high holiday of Rosh ha-Shanah.[2]

Rabbi Nathan Krass of the storied, elegant Reform Temple Emanu-El told his congregants that "it has been the avowed aim of Judaism not to divide the world into the sacred and secular, but rather to sublimate the secular life into the sacred. Religion has no quarrel with poetry or science or philosophy. . . . Hence, for the Jew, there can never be any conflict between religion and science as recently manifested in the notorious trial in Tennessee." Rabbi Maurice Harris, speaking at the newly built, grand Temple Israel, told his congregation that "evolution has tremendously advanced our conception of the eternal source behind all." Speaking to the Montefiore Congregation in the Bronx, Rabbi Jacob Katz (who also served as chaplain of Sing Sing prison), applauded the "past history of the human race [during which] our development has been characterized by a successful attempt to control nature." Rabbi Jacob Kohn of Temple Ansche Chesed preached a sermon called "The God of Truth in Science and Religion," declaring that "true religion, adoring the God of truth, will bid science Godspeed in its mighty task of conquering nature through knowledge." Further, "it is only an irrational and dogmatic religion and an irrational and dogmatic science which clash and are in conflict." Temple Beth-El's Rabbi Samuel Schulman criticized the "Literalists" who opposed science and whose "intelligence is limited. These literalists are not modern and liberal enough."[3] "The trial was certainly nothing of which, as Americans, we could be proud. By the possibility of such a trial," Schulman said, "we showed ourselves to be at least fifty years behind the age. We were made the laughing stock of Europe. We proved that while we are politically in advance of Europe's culture, spiritually we are lagging behind."[4]

Rabbi I. Mortimer Bloom of the Hebrew Tabernacle warned his parishioners:

America, once a land of light and liberty, bids fair soon to be shrouded in a pall of ignorance and illiberalism that will extinguish the torch of learning and bring back the darkness of medieval night. Contemplate the multiplying efforts and proposals to compel Bible reading in the public schools, to outlaw the teaching of organic evolution in schools and colleges, to elaborate and rigorize the Blue Laws, to introduce censorship of the stage, the cinema, the novel. What are all these but opening guns in a cunningly contrived campaign to control education, to establish statutory morality, to convert government into the secular arm of the Church and institute a state religion with religious tests for office, for suffrage, for citizenship and, ultimately, for property-owning and for residence. These grandiose schemes must be blasted, this campaign must be aborted, if America is to remain true to her traditions, if she is to keep her place among the civilized and progressive nations of the world. We must not suffer illiterates and fanatics, the back-looking elements of the community, to stop the wheels of progress, to set back the hands of time. The hour has struck for all literate, liberal and enlightened Americans to awake to the peril that menaces our Republic and, holding high the torch of civilization, to disperse the gathering darkness that threatens to blot out all that is high and holy in American life, all that has made and can still make America a beacon light to all the world.[5]

These were angry words, and fearful ones. Bloom looked to Dayton and saw that not just science was under attack. So too were secular public schools. So too was artistic freedom. So too was all that allowed Jews to live as equals in the United States. Left unchallenged, Bloom warned his congregants, the Christian attack on science could leave America a place in which Jews were denied a vote, a job, and hearth and home. As Bloom saw it, the science under attack in the Scopes trial was inseparable from the democracy, progress, and separation of church from state—all those elements of the "American way"—that had made America a land of opportunity for Jews.

Already in 1925, then, the defense of science from "literalists" and their "irrational and dogmatic religion" was seen in New York as a *Jewish* issue of the highest importance, of *high-holiday* importance. Already then, many Jews saw science as crucial to creating a public sphere—schools, universities, newspapers, voting places, places of employment—in which Jews could thrive. Already then, science mattered to America's largely immigrant Jews. In the following decades, it would come to matter more still, as many American Jews—mostly sons and daughters of immigrants—made a name for themselves in the top ranks of the nation's scientists.

It was at this time, too, that Jewish scientists became folk heroes for American Jews. An American boy celebrating his bar mitzvah between 1937 and 1951 was likely to receive, among the Kiddush cups, prayer books and other gifts, an oversized volume called *From Moses to Einstein: They All Are Jews*, by a man named Mac Davis. My father's copy sits on my shelf. The book was never out of print during these years. It made a popular coming-of-age present, as it promised both entertainment ("in this calendar of heroes and heroines, you will find melodrama, ambition, power, courage, love, despair, poverty, and all the influences that make greatness") and uplift ("to the youth these stories will reveal the possibilities to which Jews may aspire").[6]

The book opens with biographies of traditional heroes of Jewish lore, including Moses, Bar Kochba, and Maimonides. Quickly enough, it turns to modern heroes such as Abraham Schreiner, the Galician Jew who first synthesized petroleum; Siegfried Marcus, the Viennese Jew who engineered the first horseless wagon; Otto Lilienthal, the German Jew who designed the first manned aircraft; Albert Michelson, the Chicago Jew who calculated the speed of light and harvested a Nobel Prize; August von Wassermann and Paul Ehrlich, who devised a test and treatment for syphilis, respectively, and garnered Nobel prizes; Karl Landsteiner, who devised blood typing, again winning acclaim from Stockholm; and so on, until Albert Einstein.[7]

One may learn several things from *From Moses to Einstein* and its long-enduring popularity. By a third of the way through the twentieth century, Jews in the New World and the Old had registered remarkable achievements in physics, chemistry, biology, mathematics, medicine, and engineering. What's more, the scientist Jews behind these achievements were a source of swelling pride for the great majority of American Jews, who would never wear a white lab coat, titrate a solution, or integrate a function. In America in the first decades of the twentieth century, Jews had become important in science and science had become important to Jews.

Jews in Science, Science to Jews

There are many measures of the success of American Jews in science in the twentieth century. Each such measure is partial and flawed, but together they produce a persuasive picture. Prestigious prizes offer one sort of indication: 38 percent of American Nobel laureates in physics are from Jewish backgrounds, as are 42 percent of American Nobelists in physiology and medicine and 28 percent of U.S. prize winners in chemistry. Among National Medal of Science recipients, 37 percent have Jewish backgrounds.[8] American Jews are overrepresented by a factor of 15 among winners of the A. M. Turing Award for computing, and by almost as much among winners of Fields Medals in mathematics.[9]

When last studied (more than three decades ago), 26 percent of the physicists in America's best universities were Jews. One in five mathematicians, bacteriologists, biochemists, and physiologists were. Greater numbers of physicians are Jewish, and greater percentages of the faculty at leading medical schools are Jewish.[10] And one in three students at Ivy League universities is Jewish.[11]

Who's Who in American Jewry of 1938 went to great pains to include everyone who had turned a good buck in the rag trade or junk trade or retail or manufacturing or any of the other businesses in which Jews had begun to distinguish themselves, as well as every famous Jewish entertainer, athlete, newspaper reporter, writer, and rabbi. Still, about one in six of the 8,477 listings were men and women of science (over half of these physicians).[12]

Beginning in the nineteen-teens, Jews also played a growing part in bankrolling American science and medicine. Foundations established by Julius Rosenwald,[13] Ellas Sachs Plotz, Edward Filene (Century Fund), Nathan Straus, the Warburgs,[14] the Strausses, the Schiffs,[15] the Loebs, the Guggenheims,[16] the Baruchs, the Lehmans, Lewisohns, Seligmans, Laskers, Rothschilds, Littauers, Gimbels, Adlers, Fels, and on and on[17]—names known to anyone who has ever strolled on an Ivy League campus or applied for a research grant—donated hundreds of millions of dollars to hospitals for care and research, to universities to advance science, to research institutions, to planetariums, aquariums, science museums, school science enrichment programs, and more.[18]

Jewish fondness for science also became part of workaday Jewish culture. Dozens of books were written in the 1920s, '30s, and '40s cataloging the Jewish "contribution" (as they inevitably put it) to America or some-times, more expansively, to "civilization." (These books were essentially "From Moses to Einstein" for adults.) Beginning in New York in 1927, the Jewish Academy of Arts and Sciences gathered scientists, doctors, and men of letters to debate scientific and scholarly matters—and as far as I can tell, to distribute awards, plaques, and commendations one to the other in sur-prisingly great numbers—with the notion of advancing science and, no less, advancing the notion that *Jews* are advancing science. When the Federation for the Support of Jewish Philanthropic Societies of New York petitioned the WPA Art Project to provide a mural for the Hebrew Orphan Asylum, the seven-panel work they received was called *Theory and Practice*. William Karp, the Jewish painter, depicted in the center-wall panel "a heroic figure of a man, surrounded by crumbling structures and new and modern buildings, holding a blueprint for the future in his hands," suggesting "cooperation between the practical man and the man of science."[19] Jewish institutions of all sorts lavished honors and prizes upon Jewish scientists, including some of arcane accomplishment.[20]

Books of popular science became mainstays of New York's Yiddish pub-lishing houses.[21] In 1915 the Folk-Education Publishing House issued two Yiddish books in a single binding: *Darwinism and His Theory: A Discus-sion of Popular Science* and *Man & Woman: Living among the Animals*.[22] The Arbeiter Ring, or Workers Circle, took it upon itself to publish the basic handbooks that its members would need as they acclimated themselves to New York life. By 1918, they had issued these Yiddish titles: *Botany, Phys-ics, Hygiene, Trade-Unionism, Socialism from a Social-Revolutionary Point of View, Astronomy, Geography, Political Economy,* and *Physiol-ogy*.[23] In 1920 Solomon Herbert's *First Principles of Evolution* was trans-lated into Yiddish and published simply as *Evolution*.[24] Later, the Popular Scientific Library published G. A. Guryev's *Darwinism and Atheism*.[25]

And it wasn't just books. Synagogue Sisterhoods organized field trips to science museums and planetariums.[26] Plus, as we have seen, rabbis regularly praised science from the pulpit, often lauding Judaism for its embrace of the same modern knowledge that so irked fundamentalist Protestants.

THE MAKING OF AMERICAN JEWRY

The story of how science gained such a hold across so wide a swath of early twentieth-century American Jewry—anarchists, socialists, and capitalists; atheists, Reform, Conservative, and Orthodox; poor and rich; established and immigrant—is complicated and inseparable from the exquisite com-plexity of the social conditions of Jews in those years.[27] Between 1881 and

1914, two million eastern European Jews settled in the United States. The scale of this immigration is difficult to absorb from today's remove. In 1880, New York had the largest Jewish population in America, with about 80,000 Jews, most of them descendants of German Jews who arrived in the years immediately after 1848. Thirty-four years later, that number had grown to 1.4 million, an increase of 1,750 percent. From the beginning, this greatly accelerated immigration was a cause for alarm among the German Jews already settled in America. It is easy to understand why. The new arrivals tended to be poor and poorly educated. They were, by the standards of the earlier, German immigrants, uncultured primitives. The *Jewish Messenger*, a New York weekly, described them as a "mass of illiterate, uncouth, mainly superstitious foreigners." This evaluation was undoubtedly amplified by the insecurities of the earlier, German Jewish immigrants, who saw in the new ones what they feared (and with good reason) Protestant Americans saw in *them*. As steamship after steamship arrived, the Jews on shore couldn't help but worry that the Jews stumbling down the gangplanks were going to make America a less sympathetic place for *all* Jews.[28]

This was not an idle worry. The steady arrival of so many eastern European Jews did occasion new, alarming strains of American anti-Semitism. Though German Jewish notables had trouble entering polite Protestant society, they still enjoyed remarkable rapport with leading Christian Americans. There were many signs of this rapport, as historian John Higham has written: "Protestant ministers and Reform rabbis frequently exchanged pulpits. Rising Jewish capitalists joined in general community affairs and built lavish homes in the most exclusive neighborhoods. The traditional American image of the Jew as a constructive economic force—a model of commercial enterprise, energy and integrity—still provided material for popular orators and storytellers."[29]

This benign view of Jews as benevolent exotics (which, it must be added, rarely existed in pure form, because it was undercut by vague suspicions that Shakespeare's Shylock also captured some fundamental truth about Jewish character) did not wear well in the rising tide of *Ostjude* immigration. Fear of the conniving "international Jew" gained currency in patrician circles, especially in New York. "In a society of Jews and brokers," Henry Adams lamented in 1893, "I have no place."[30] Henry James returned to the city in 1904, after twenty years in Europe, only to find that in his absence it had fallen to the "Hebrew Conquest of New York."[31] And as a newly invigorated distaste for Jews spread among the parlors of East Coast society, a new anti-Semitism took hold downtown as well, among poor immigrant Christian laborers. An 1895 letter to the *New York Sun* signed by someone who called himself "workingman" offered that either the city's Jews would soon control everything or they would be dead; there was no middle ground.[32]

Bigotry was not the biggest worry among new Jewish immigrants. More consuming was the cruelty of their circumstances. In New York, Jews found poverty rivaling the worst of what they had suffered in Europe. By 1893, the Lower East Side of Manhattan, which became the most conspicuous of the city's "Hebrew" neighborhoods, was (as William Dean Howells, the editor of *Harper's Magazine*, put it) "more densely populated than any other area in the world, or at least in Christendom, for within a square mile there are more than three hundred and fifty thousand men, women and children."[33] A ramshackle tenement with sixteen tiny apartments housed more than 200 Jews. "They crowd as do no other living beings to save space, which is rent, and where they go they make slums," wrote Jacob Riis, the author of an explosive exposé of tenement life called *How the Other Half Lives*, and "the poverty they have brought us is black and bitter."[34] Though they later came to be viewed with nostalgia, as these things do, the bursting Jewish neighborhoods around the start of the twentieth century were places of deprivation and anguish.

They were also places of energy and ferment. Howells was bemused by the odd intensity of the Hebrew economy of New York: "These people were desperately poor, yet they preyed upon one another in their commerce, as if they could be enriched by selling dear or buying cheap. So far as I could see they would only impoverish each other more and more, but they trafficked as eagerly as if there were wealth in every bargain. The sidewalks and the roadways were thronged with peddlers and purchasers."[35] Riis also registered the vigor of Jewish ambition: "As to the poverty, they brought us boundless energy and industry to overcome it. Their slums are offensive, but unlike those of other less energetic races, they are not hopeless unless walled in and made so on the old world plan. They do not rot in their slum, but rising pull it up after them. Nothing stagnates where the Jews are."[36]

And it was not just in matters of livelihood that the immigrant neighborhoods were dynamic. Though they lacked the sorts of underwriters that kept midtown and uptown culture afloat, the downtown ghettos produced many varied entertainments. In 1884, an actor named Moshe Silberman led his Romanian acting troupe to New York, where they began performing Yiddish drama. Acting troupes were joined by cabarets. By early in the twentieth century, Jewish neighborhoods hosted more nickelodeon movie parlors, by a wide margin, than any other neighborhoods in the city. Coffee shops became loud, smoky community centers, housing political meetings and poetry readings. Dozens of Yiddish papers, magazines, journals, and broadsheets came into existence, folded, and were replaced.

Immigrant neighborhoods in New York and elsewhere were also political hothouses in which esoteric ideologies speciated and thrived. Already in 1883, a young Abraham Cahan (who went on to be the legendary founding editor of the Yiddish daily *Forward*) held hundreds enrapt as he lectured on Friday nights on Marx in Yiddish (in one instance calling, with a flourish, for his audience to join him in storming Fifth Avenue with axes and swords). The next years saw the founding and rapid growth of Jewish labor unions and Jewish divisions of socialist political parties and anarchist alliances.[37] (Shalom Aleichem, the brilliant Yiddish writer who moved to New York from Ukraine in 1905, began a story about his new home: "It seems to me that there is no better thing in the world than a strike.")[38] With time, these were joined by Bundists, Zionists, and return-to-the-land schemers of utopian sorts (which produced egalitarian communal farms in New Jersey and Oregon, the latter called New Odessa) and less utopian sorts (which trained immigrants for agriculture and helped them settle upstate and, once again, in New Jersey).[39] In 1914, Lower East Side Jews swept socialist Meyer London into Congress. When he died twelve years later, half a million of them attended his funeral; as the *New York Times* reported it, a "solid mass of people" wept in a "pageant of sorrow."[40]

There were many reasons for the tumult and ferment in the neighborhoods of the eastern European Jewish immigrants. Manic energy and creativity were part of the experience of other immigrants to America and may—like high school experiments where broken chemical bonds produce extraordinary heat—just be the natural result of ethnics reconstituting their lives on foreign shores. Still, even if not altogether unique, the experience of the eastern Jews was extreme in its intensity and results. For instance, Jews went from being the poorest subculture in America to the richest in a span of less than three generations, a remarkable and incomparable rise.

The vitality of the Jewish slums owed something to the peculiar nature of the Jewish immigrants themselves, who were different in some ways from the other immigrants they brushed against at Ellis Island. For one thing, even the most crudely educated among them were typically literate, and among the immigrants were a good number who had received a superior *gymnasium* education in Europe. Even for the more *lumpen* of the proletariat, who did not come with a working knowledge of Pushkin and Marx, the crowded neighborhoods of New York, Boston, Philadelphia, Chicago, and Detroit were manageable in part because these newcomers drew on skills that Jews had exercised in Europe. Irish, Italian, and Russian peasants who made their way to America often had to learn to negotiate urban life and livelihood from scratch; their new city lives bore little resemblance to the agrarian lives they had left behind. But Jews had lived as peddlers, shopkeepers, and craftsmen for generations in Europe. The transition from the slums of Kiev to the slums of Canal Street was not as demanding for Jews as the trip from homesteads near Belfast to the slums of South Boston was for the Irish.

The vitality of turn-of-the-century Jewish life owed something as well—though just how much, it is impossible to say—to an enduring ambivalence about America itself. Many immigrants saw in America a promise of fair treatment and unencumbered opportunity. Abraham Cahan described in the *Atlantic Monthly* how, after the first wave of pogroms, a small group of secular students made their way to the pulpit of a Kiev synagogue, where one of them implored the weeping religious Jews that filled the benches: "There is no hope for Israel in Russia. The salvation of the downtrodden people lies in other parts,—in a land beyond the seas, which knows no distinction of race or faith, which is a mother to Jew and Gentile alike. In the great republic is our redemption from the brutalities and ignominies to which we are subjected in this our birthplace. In America we shall find rest; the stars and stripes will wave over the true home of our people. To America, brethren! To America!"[41]

"MAKING IT" IN AMERICA

It did not take long for immigrants to realize that the promise of a land that "knows no distinction of race or faith" and offers "redemption from . . . brutalities and ignominies" would not be fully fulfilled on their new shores. The *New York Times*, the *World*, and the *Herald-Tribune* ran want ads stipulating that "Christians only" or "Protestants Only" need apply. Clubs remained "restricted," often barring Jews even as guests. University quotas placed ceilings on Jewish attendance. Beginning in 1915, the resurgent Klan made anti-Semitism an inseparable aspect of its race-based worldview. "The Protocols of the Elders of Zion," translated into English and serialized by Henry Ford beginning in 1920 as *The International Jew: The World's Foremost Problem*, eventually sold upward of half a million copies as a book. Immigration laws were revised to limit, and eventually to end, further Jewish immigration to America. These restrictions were backed by new scientific data testifying to the inferiority of Jewish (as well as some other) stock, the weakness of Jewish character, and the inconstancy of Jewish patriotism.

American Jewry in the first decades of the twentieth century came to

rest atop significant fault lines under mounting pressures. The public leaders of American Jewry—the great philanthropists, the politicians, the heads
of emerging Jewish organizations, the successful bankers, businessmen,
newspaper owners, and so forth—were almost all descendants of German
Jews who had immigrated several generations earlier. They were ambivalent
about the great mass of eastern European Jews they ostensibly led (and the
reverse was equally true). The religious orthodoxy of many of the new immigrants; the political extremism of a loud minority of them; their stubborn
insistence on speaking, writing, and reading Yiddish; their dingy presenta-
tion and unremitting poverty—all these things left the "greenhorns" (as the
new immigrants were disparagingly known) uncomfortably un-American.
For some of the German-Jewish leaders, civilizing and Americanizing the
inhabitants of neighborhoods like the Lower East Side of New York became
their highest priority.

Schemes were organized to distribute Jews in small numbers to more au-
thentically American locales; an organization called the Jewish Immigrants'
Information Bureau diverted 10,000 immigrants to Galveston, Texas, between 1907 and 1914, for instance. And charitable organizations devised
programs operating within Jewish neighborhoods to speed the assimilation
of immigrants. The National Council of Jewish Women sponsored Jewish
settlement houses around the country, where immigrant children might be
introduced to the Boy Scouts, learn prairie folk songs, receive instruction
in nutrition, or otherwise acquire the basic skills and concepts of American
life. These schemes and programs were the product of real concern for the
welfare of newly arrived Jews facing a harsh reality. But they were also a
product of anxiety on the part of uptown Jews that the masses of downtown
Jews might diminish their own hard-won status as Americans.[42]

For their part, most downtown Jews also wished to assimilate, to "make
it" in America. "Making it" was the great recurring theme of Jewish aspiration in the first half of the twentieth century. It was a broad notion, one
that meant more than just striking it rich. It meant acquiring acceptance
and esteem. It meant making it economically, making it culturally, making
it socially, and making it politically.[43] Yet, among newer immigrants and
their children, just how they might succeed in America was a matter of some
uncertainty and ambivalence. Most were zealously patriotic. But many bal-
anced a deep, almost religious belief in the ideals of America—a visceral
patriotism born of gratitude and fortified by the constant stream of horrid
news about the fate of European Jewry—against their recognition that these
ideals were not uniformly applied and that, for all the happy rhetoric about
equal opportunity for all, being a Jew remained a liability in professions,

business, and society. They also found that they had to balance their feeling of being real Americans (which increased exponentially from immigrants to their American-born children) with an irreducible insecurity and the feeling of needing to *demonstrate*—to others, but no less to themselves—that they were real, contributing Americans.

For decades at the start of the twentieth century, American Jews excelled at boxing and then basketball. During the years of Jewish dominance of these sports, star Jewish athletes were valorized much like scientists would soon be. Joe Bernstein, a champion featherweight who fought at the start of the twentieth century, was known as "The Pride of the Ghetto." After Bernstein retired, boxer Louis Wallach took up the mantle as "The Pride of the Ghetto"; and in turn, when Wallach hung up his gloves, Ruby Goldstein fought as "The Jewel of the Ghetto."[44] Goldstein gave way in the pantheon of Jewish boxers to Benny Leonard, lightweight champion of the world, who wore a Star of David on his trunks and declared that in the ring he meted out comeuppance for all the thugs who had ever bloodied the nose of a Jew.[45] College and professional basketball stars—Max "Marty" Friedman, Nat Holman, Sonny Hertzberg, and Red Holzman, to name just a few—were also sources of pride and pulsing self-satisfaction.[46] Jewish papers reported their exploits with verve, and Jewish intellectuals theorized about what their success had to say about Jewish character, and about the Jewish contribution to advancing the culture of America. Compendia were published of the exploits of the best American Jewish athletes. In all, these sportsmen were offered, with swells of pride, as a demonstration that the calumnies of those who disparaged Jews and their contributions to society were without basis in fact.[47] In this regard, physicists were not so different from pugilists; both were a source of anything-you-can-do, -a-Jew-can-do-better pride and an apologetic demonstration of the civic worth of a Jewish minority.[48]

But all this explains only the simplest link between Jewish success in science and the broad Jewish embrace of science. In a more profound way, science was embraced by Jews because it was taken to be a harbinger and herald—at once cause and effect—of a new social and political ideology that well matched the aspirations of Jews throughout Europe, America, and Palestine in the first decades of the twentieth century.

SCIENCE AND THE MAKING OF MERITOCRACY

In the minds of many Jews, and with good reason, science was associated with the ideal of secular, liberal democracy. Science as it came to be practiced in the twentieth century offered exactly what many Jews wanted most:

a vocation where what mattered was what you accomplished, not where you came from. The ideal of science, as many Jews came to understand it, was of an endeavor blind to the background of those who pursued it. As such, science seemed to offer Jews a tool, unparalleled in its effectiveness, for entering into the non-Jewish societies where they found themselves. It seemed to promise a seat at the table, and sometimes the supremely elegant table of the Royal Society, or the National Academy of Arts and Sciences, or the Academy of Sciences of the USSR.

It is easy from today's remove to forget how novel, radical even, this notion of blind meritocracy was. Near the start of the twentieth century, in the United States, for example, Jews were customarily barred from many traditionally prestigious occupations. Top-tier law firms took on few Jews, and those they did rarely made partner. One contemporary estimate ran like this: "In general office work, about ninety percent of the jobs available in New York are barred to Jews."[49] Though this estimate was almost certainly an exaggeration, it felt accurate enough to Jews of the time, for whom discrimination of one sort or another was a fact of daily life. Discrimination was documented in a great variety of fields, including banking, insurance, brokering, advertising, journalism, upscale retail, restaurants, and many more. Some fields had come to be dominated in New York by Jews, fields such as the garment and fur trade, textiles, jewelry, accounting, drug and tobacco stores, as well as entertainment. But fields in which Jews and Christians worked side-by-side as equals with equal opportunities—these were more rare.

Indeed, at the start of the twentieth century, Jews who wished to enter even the sciences also met with resistance. In 1908, *Science* magazine could still complain that "democracy has not until lately joined itself with the educated classes for the promotion of scholarship." "Scholars and scholarship," it noted, were still "the allies of aristocracy."[50] But already this state of affairs was changing quickly in science in the United States and, indeed, throughout most of the West.

In the United States, Andrew Carnegie helped spur this change when he set out to reform both university education and the scientific establishment. The Carnegie Foundation for the Advancement of Teaching offered fat grants only to those universities willing to "modernize": to relinquish Christian affiliations, eliminate classical language requirements, and beef up science instruction, especially lab instruction. The Carnegie Institution of Washington, established in 1902, began with a $10 million gift (which was greater than the research endowment of all American universities combined), and was charged with encouraging "investigation, research and discovery" on a purely "scientific" basis. An important aim of the institute, Carnegie said, was "to discover the exceptional man . . . whenever and wherever found, inside or outside of schools, and enable him to make the work for which he seems specially designed his life work."[51] This was one short year after the Rockefeller Foundation agreed to launch the Rockefeller Institute to develop a scientific understanding of "the nature and causes of disease and the methods of its treatment." These institutions, and the many others they spawned or resembled, promised to remove the traditionalism and conservatism that stifled innovation and to replace them with a new sort of science the value of which was measured in the data produced, not in the prestige of the men who produced these data.

Owing to such reformist efforts as well as to many other factors—growing public, government-run science; changes in the universities; the spread of new European models of scholarship like that embodied in the Kaiser Wilhelm Institutes; the rise of Taylorism and scientized administration; a rise in corporate research and development; and more—the infrastructure of science as we know it today took new shape during these years.[52] Blind refereeing came into common use in judging articles submitted for publication and proposals submitted for research grants.[53] While discrimination persisted at almost every stage—training, employment, advancement, public recognition—scientists and scientific institutions seemed to be sidestepping with growing success the exclusionary leather-wingchair-and-brandy-snifter Protestantism that had stymied Jews in other areas. The most important scientific institutions in America were new, without stultifying elite traditions and traditional elites.

For Jews drawn to it, there was evidence that science offered opportunities not found outside the laboratory. In 1913 Jacques Loeb, a French-Jewish biologist who stunned the world when he induced parthenogenesis by chemical stimulus, "creating artificial life" (as the awestruck press had it), was kept out of New York's exclusive and exclusionary Century Club because he was a Jew. J. McKeen Cattell, the Columbia psychology professor who proposed Loeb's membership, was outraged at the old-fashioned injustice, quite out of step with modern times. He wrote to the club's admission committee: "Dr. Loeb is one of the greatest scientific men of the world, one of the great men of the world. There are not in the United States twenty scientific men more eminent; there are not among the members of the Century Club twenty men more distinguished. . . . You have rejected Dr. Loeb, even though you would not dare admit it, because he descends from and has the traits of that race which before and since the birth of

Christ has supplied in proportion to its numbers more men of genius, the Greeks only excepted, than any other."[54]

Loeb himself scoffed at the old-fashioned "snobocracy" that had rejected him, predicting that the old elites were likely to "degenerate" and confident in the esteem in which he was held by the meritocratic men of science.[55] The Rockefeller Institute boasted of having the great biologist on its faculty. In the eyes of many Jews (and not only Jews), it said something about science, just as it said something about Jews, that Loeb's work as a scientist was so highly respected. It also said something about science and Jews that Sinclair Lewis made a dashing hero of Loeb as the fictional Max Gottieb in his Pulitzer Prize–winning novel, *Arrowsmith*; that John Ford did the same in his popular movie of the book; and that Loeb became what one historian called "a central symbol of pure science in America during the years between the wars."[56]

It said something, too, that another American Jew, Simon Flexner, became the founding director of the Rockefeller Institute, and that his older brother Abraham Flexner founded the Institute for Advanced Study in Princeton, bringing Einstein to the United States. Einstein himself was an icon of incomparable fascination, who for decades provided ubiquitous proof of how far a Jew could advance in science, and how far science could take a Jew—to the tables of heads of state and the palaces of kings. When, in 1921, Einstein toured the United States with Chaim Weizmann, then president of the World Zionist Organization and later Israel's first president, and Menachem Mendel Ussishkin, secretary of the Zionist Congress, he was met with adulation never before accorded a Jew on American soil. The ship they arrived on, the S.S. *Rotterdam*, was met by "vast throngs of New York Jews [who] turned out to greet Professor Chaim Weizmann, discoverer of T.N.T. and president of the World Zionist Organization, and Professor Albert Einstein, famous savant."[57]

But the enthusiasm for Einstein was hardly limited to Jews. After a brief political tussle about the probity of his pacifism, the New York State Legislature conferred upon Einstein its highest honor, the "Freedom of the State of New York." He was invited to lecture at the finest universities in the country, agreeing to speak at Harvard, Columbia, Princeton, and the University of Chicago, among others. His eight-week visit was covered in more than 160 articles and essays in the best newspapers and magazines in the country. William Carlos Williams wrote a poem, "St. Francis Einstein of the Daffodils," to mark " the first visit of Professor Einstein to the United States in the spring of 1921":

Sing of Einstein's
Yiddishe peachtrees, sing of
Sleep among the cherryblossoms.
Sing of wise newspapers
That quote the great mathematician:
A little touch of
Einstein in the night.[58]

The appreciation that America's Jews showed for Einstein could be divided into two familiar sorts. One was for his scientific achievements themselves, which, though few grasped them in accurate detail, all knew to be brilliant and important. The other was for demonstrating to everyone else that Jews were worthy of respect. When Abba Hillel Silver, the renowned Cleveland rabbi and Zionist leader, took to the dais before 5,000 at New York's Metropolitan Opera House, with hundreds of dignitaries crowded behind him on the stage, he said: "We greet, all of us, that man, that intellectual Titan, who has again given evidence through his labors and his achievements of the intellectual leadership of the sons and daughters of Israel throughout the world, Albert Einstein."[59]

Einstein's 1921 visit was treated as, among other things, great evidence of what Jews had what to offer, throughout the world. When, in 1933, Einstein agreed to take up residence in Princeton and employment at the Institute for Advanced Study, he became a potent symbol for many American Jews of what Jews could achieve if only given a chance.

Responding to an essay in the *New Republic* charging that Einstein found only one or two Americans who understood his theory, the towering City College philosopher Morris Raphael Cohen wrote that "Einstein has a most generous admiration for these American men of science who, like Michelson, Milliken or Jacques Loeb, are making important contributions to the advancement of human knowledge."[60] It would not be lost on readers of the *New Republic* that two of the three scientists, like Cohen himself and like Einstein, whom he defended, were famously Jewish.

Scientific Democracy: The Ethos of Science

The sudden success and prominence of Jews in American science meant something to many American Jews. It meant, of course, that the sciences were a vocation that Jews *could* enter, and one in which they would be allowed to rise to the highest levels. Jews might be restricted from social clubs but not from the citadels of science. Science, for American Jews, was a career option. But this practical virtue—shared by many other occupations—was not the soul of the attraction of science. Science also seemed to embody a set of ideals that many Jews, seeking to find their place in America, embraced. Among these ideals were meritocracy and the democratic notion that in the public sphere every person ought to have a voice, regardless of background or religion. Another of these ideals was the notion of *blind* evaluation—that the worth of ideas cannot be measured by reference to the Social Register.

Views like these turn up over and over in the writing of Jews of the time. In 1928, when Morris Raphael Cohen was invited to give the annual address to the Judaeans, a society of worthies (including some of the most famous Jewish judges and professors of the day) that aimed to advance the "intellectual and spiritual interest of Jews," the topic he chose was "The Jew in Science":

> The Jews have since the end of the eighteenth century been a people on the move. They have participated in the rapid expansion of Europe. This expansion has broken down the walls of the ghetto. And the result is that the Jews are in a state of flux or transition. A gas that has just been taken away from some compound and liberated combines more readily with new elements. So the Jews, being in a nascent or transition stage, are eager, and have the zest or spirit of adventure essential for modern science. Breaking away from their old routines, the Jews are in large measure spiritual explorers, ready to enter new movements. It is thus that they entered the field of science.[61]

For Cohen, who wrote elsewhere that "the first fact we Jews in the United States must never forget is that we are American citizens,"[62] science was an avenue for Jews to break free from the cables of their past ghetto identities and embrace new identities as people and, in the event, liberal Americans. Just how science could do that was described in an influential book by the London mathematician Hyman Levy, published first in London and then in New York in 1933, and going through several editions in both places: "The ideal of scientific knowledge is to reach a system capable of being isolated from the subjective world of every individual member of the human race. It would be a set of statements acceptable to *all*. . . . The formulations of science . . . are statements *invariant* with respect to the individual."[63]

The Jewish American sociologist of science Robert K. Merton (whose given name was Meyer R. Schkolnick, although he did not speak publicly of his Judaism until his eighth decade) quoted Levy in a moving essay called "Science and the Social Order," which appeared in the journal *Philosophy of Science* in 1938. Merton wrote that "the sentiments embodied in the ethos of science [are] characterized by such terms as intellectual honesty, integrity, organized skepticism, disinterestedness, impersonality. . . . It is a basic assumption of modern science that scientific propositions 'are invariant with respect to the individual' and groups. . . . One sentiment which is assimilated by the scientist from the very outset of his training pertains to the purity of science. Science must not suffer itself to become the handmaiden of theology or economy or state."[64]

In the same year, Franz Boas, the great German-Jewish émigré anthropologist, published a "Manifesto on Freedom of Science" bearing the signatures of 1,284 prominent American scientists. The document, which Boas had painstakingly produced and advanced for years, was an alarmed response to Nazi repression of Jewish scientists and suppression of their work: "American scientists, trained in a tradition of intellectual freedom, hold fast to their conviction that . . . 'Science is wholly independent of national boundaries and races and creeds and can flourish only where there is peace and intellectual freedom.' . . . We firmly believe that in the present historical epoch, democracy alone can preserve intellectual freedom. Any attack upon freedom of thought in one sphere, even as non political a sphere as theoretical physics, is in effect an attack on democracy itself."[65]

Boas, who was eighty years old at the time, devoted himself with remarkable energy to protecting democracy by defending science, and protecting science by defending democracy. He traveled and lectured on the subject, wrote letters to colleagues, lobbied politicians, and continued to write for academic journals and popular magazines alike. A *Forum and Century* magazine article called "Race Prejudice from the Scientist's Angle" offers a good portrayal of the mix of science and politics that drove Boas into his ninth decade:

> There are Nazi physicists who repudiate relativity because it was formulated by Einstein, a Jew. Professor Bieberbach of the University of Berlin insists that there is a Jewish approach to mathematics. And Professor Lenard, Nobel Prize winner, proclaims his deep conviction that only the "Nordics" have made valuable contributions to science.
>
> We are not free from these tendencies in the United States. There is a rising tide of race prejudice and especially of antisemitism and anti-Catholicism. The obvious remedy is education—teaching the indisputable fact that color of skin, class, religious belief, geographical or national origins are no tests of social adaptability. Yet in the face of this need, we find schools and colleges limiting the number of Jewish teachers and students. It is time to restate the beliefs of the founders of this nation and drive home again the democratic principle that a citizen be judged solely

by the readiness with which he fits himself into the social structure and by the value of his contributions to the country's development.[66]

Boas' view of science was not uniquely Jewish; a great number of the scientists who signed his manifesto were Jews, but a number greater still were not.[67] By the time the guns of the Second World War fell silent, it was very nearly an article of faith among all scientists, Jewish and not.

Historian Andrew Jewett has called those who hold such views "scientific democrats," and has described how their influence grew and endured from the Civil War to the Cold War. "From the 1860s to the 1960s," Jewett wrote, "scientific thinkers in the United States repeatedly insisted that science did imply certain values—in fact, exactly those values needed to sustain the cultural foundations of American democracy." Scientific democrats were persuaded that the values embodied and advanced by science were sufficient to replace Christianity as a binding adhesive for American citizens, allowing them to "bring their institutions into line with their needs, and thus sustain self-governance, without converging on a shared Protestant worldview." Science could achieve this feat "by shaping their moral character, normative commitments, and discursive practices."[68] For those who adopted this view (and as Jewett has shown, there were many, in the human sciences as well as the physical and biological sciences), the social and political impact of science could be enormous:

> Science, in short, promised thoroughgoing cultural change, rather than simply the augmentation of the nation's knowledge base. In this understanding, science denoted a personal orientation, not just a body of knowledge or a set of institutions. Being scientific meant much more than simply using empirical methods; it meant behaving in accordance with specific ethical tenets or exhibiting particular ethical virtues. It entailed a mode of speaking, a form of interpersonal relations, even a comprehensive way of life. Scientific democrats portrayed the scientific enterprise— the whole complex of practices, institutions, knowledge claims, and persons—as a concrete manifestation of an underlying ethical orientation that was perfectly suited to the needs of a modern democracy.[69]

In Jewett's account of the scientific democrats one finds many Jews, their numbers growing as the decades of his account advance. Among American Jews, some form or another of scientific democracy was all but universal.[70] This view of science and its place in American life was championed in the speeches of Jewish scientists—which is perhaps unsurprising—but it is valorized no less in the last wills and testaments of Jewish philanthropists, the essays of Jewish philosophers, the addresses of Jewish politicians, the news article of Jewish journalists, the plotlines of Jewish novelists. It was a view shared by German Jews and *Ostjuden*, by Zionists and anti-Zionists, by businessmen and union organizers, by Bundists and assimilationists, by Reform, Conservative, and Orthodox Jews. Save hatred for the Nazis, no other attitude enlisted such broad consensus in this famously fractious community in its most fraught and divided decades.

This was so, in part at least, because this ethos of science fit snuggly with the broader aims of the largely immigrant Jewish population in America, eager to establish a place for themselves in their new home. Journalist and bestselling author Harry Golden wrote in his 1950 *Jews in American History: Their Contribution to the United States of America* that

> science like art knows no racial or geographic compartments since the ultimate aim is the service of mankind. Science does not set apart groups nor does it know distinctions. Science's only concern is for the furtherance of human welfare. Jews like every other nationality have woven threads into the warp and woof of the pattern. If you attempt to remove a thread, the pattern will unravel. Every sphere of life is of a similar pattern and similar nature. We should carry the portent of its message into our daily lives. What affects one thread affects the pattern. No one group has a monopoly on anything, everything belongs to humanity.[71]

Golden saw it like this: science, because it is indifferent to the beliefs and ideologies of scientists, carries a message that matters to Jews trying to make it in America and elsewhere in the West, that "no one group has a monopoly on anything," and that everything belongs to everyone. This message, Golden and many others thought, was what made America work for Jews.

Secular Schools, Science, and a Civil Society

Many American Jews in the generations before and after the Second World War looked to public schools to ensure that this message took root. For them, state schools promised a way out of poverty and a way into American boardrooms and clubrooms. To provide these things, however, public schools had to be of a certain character. They had to be hospitable to Jews, and they had to train these Jews in the ways of America. But perhaps their most important job was to advance the ideal described by Golden, Boas, Merton, and so many others, not just among Jews but among Christian children as well, of an America that judged people by their character and achievements, not by their creed and their ancestors. Jews sought from public schools not just the intellectual and cultural training needed by their children so that they could be hired into good middle-class jobs. Jews also sought from public schools the ideological indoctrination needed by the children of their Protestant and, to a lesser degree, Catholic neighbors so that they would one day be willing to hire Jews. Put differently, Jews looked to the public schools to advance a nonsectarian ideal of American public life, in which individuals would be judged by their mettle and not by their lineage, and certainly not by their beliefs. They looked to public schools not simply to remain mum about all matters of religion, but to advance the ideal that religion ought not have a place in public life.[72]

This ideal was far from universally accepted in the years between the world wars, when the greatest number Jewish immigrants struggled to establish themselves. In 1925 (the year of the Scopes trial) alone, there were initiatives in New York, New Jersey, California, Colorado, Ohio, Wisconsin, and Tennessee to beef up the religious character—in some cases, the Protestant sectarian character—of the public schools. The means varied—mandating or sanctioning religious study in the schools, banning evolution, introducing daily chapel, mandating recitation of the Lord's Prayer, introducing formal Bible study, and more—but in the eyes of contemporary Jews, the end was always the same: to prevent public schools from creating the sort of nonsectarian public culture they avidly sought. Jewish leaders fought these religious initiatives with vigor and they were uncharacteristically united in doing so; Reform, Conservative, Orthodox, and secular leaders alike joined in the opposition.

Louis Newman, who in 1925 edited a book entitled *The Sectarian Invasion of Our Public Schools*, wrote: "The shrine of our common Americanism is the public school. Our children enter it not as Catholic, Protestant, Jew or unbeliever, but as Americans all. Nothing should ever be done, as the Bible amendment seeks to do, to split the student community into warring sectarian camps. The American common school is our greatest contribution to the progress of education; we must keep it intact and inviolate."[73] Rabbi Samuel Schulman, the charismatic rabbi of New York's Temple Beth-El, sermonized with urgency in 1925: "It is best for the American people that the child in the public school should not have either his religious affiliation, as inherited from his parents, or his racial origin, in any way emphasized. In the school, American unity should, above all, be insisted upon. In the school, the teacher should know only American children."[74] The United Orthodox Jewish Congregations of Cleveland issued a statement protesting "solemnly against the . . . attempts to divide the children of the public schools into separate racial, national or religious groups in which they belong."[75] And

Rabbi Stephen Wise, chairman of the American Jewish Congress, dryly remarked that "when the Church infringed on the State the Jews would be the first to suffer."[76]

Wise's observation, a weary acknowledgment of the fragility of the status of Jews in America, was perhaps an inverted image of the grandiose idealism of his rabbinic colleagues like Newman, who described the public schools as a "shrine of our common Americanism." Wise said what some Jews of his day did not wish to admit in public, that advancing in and through the public schools an ideal of religious indifference was good for Jews, and that to do otherwise would be bad for Jews. Idealism and self-interest are not always at odds, and even if Jews benefited most from the exclusion of all religious expression from the public schools, this does not mean that the ideals they trumpeted in doing so were hollow. It is often the case, as it was here, that high-minded ideals are enmeshed in a ravel of parochial interests and concerns.[77]

Until the Second World War, Jewish advocacy of a strict ban on any religious forms in public schools took the shape of public proclamations and quickly drafted resolutions that carried no clout. Indeed, in the decades between the wars, religious sentiments and ceremonies were common in public schools. This was true even in New York City, where in some schools Jews found themselves in the majority and in others a sizable minority, and where Christmas pageants and daily prayer were typically in the curriculum. And of course these things were taken for granted in most of the rest of the country. This state of affairs began to change only in 1941, when a ragtag coalition of civic groups—the League of Women Voters, the New Jersey Taxpayers Association, the ACLU, and the Junior Order of United American Mechanics among them—challenged a New Jersey law authorizing state subsidies for busing children to parochial schools. By 1943, the New Jersey Supreme Court had struck down the subsidies in the case now known as *Everson v. Board of Education*. This ruling was reversed by the New Jersey Court of Errors and Appeals in 1945, and two years later the reversal was upheld by the United States Supreme Court. In his majority opinion, however, Justice Hugo Black wrote that the busing subsidy was not itself a religious matter. He went on, writing that the establishment clause of the Constitution erected "a wall of separation between church and state [that] . . . must be kept high and impregnable."[78] Advocates of the separation of church and state had technically lost *Everson*, but they walked away with a greater victory than they might earlier have dared hope.

The American Jewish Congress, the American Jewish Committee, and the Anti-Defamation League had declined to file amici briefs supporting the plaintiffs in *Everson*, but they were electrified by its outcome. Above all, Leo Pfeffer, staff attorney of the American Jewish Congress, saw *Everson* as proof that the courts could and should be used to remove all vestiges of "sectarianism" from the public schools. Under Pfeffer's leadership, the American Jewish Congress, the American Jewish Committee, and the ADL became important partners in a twenty-five-year campaign to enlarge the "wall of separation between church and state." They were not alone, of course; the ACLU was at the forefront of this effort from the start, and they were joined by many smaller, ad hoc groups like Protestants and Other Americans United for the Separation of Church and State. Still, Jewish organizations were at the center of the struggle to diminish the role of religion in the schools and other public institutions, and they provided increasingly experienced and confident legal assistance. The struggle to diminish the role of religion in the schools and other public institutions found its place at the heart of the identity of important American Jewish organizations and, increasingly, at the heart of the personal identity of many American Jews.[79]

In the fullness of time, many issues became embroiled in the church-state conflicts surrounding the schools: funding, busing, prayer, holiday celebration and decoration, Bible study, and creation/evolution. The last issue—the challenge to the place of evolution in the public biology curriculum—has had especially lasting significance because science itself was so closely associated in the minds of many Jews with the ideal of a civil society blind to creed, race, and background. It is for this reason that American Jews have opposed the teaching of creation and intelligent design in public schools with a unity of voice and near-consensus greater than that of any other ethnic or religious group on American soil.[80]

One man for whom biology came to be linked closely with the ideal of a meritocratic civil society was Benjamin Gruenberg, who came as a nine-year-old to New York from Russia in 1883, and rose to head the biology and science departments of the New York City school system from 1902 to 1920, the height of Jewish immigration. In 1911, he completed his doctorate in genetics at Columbia, working under the renowned Thomas Hunt Morgan. Gruenberg wrote two of the most popular biology textbooks of his day, a handful of other popular books about science, and a great many academic and popular essays. He served as director of the American Association for Medical Progress; as education director of the Urban Motion Picture Industries, producing films on science for schools; and as educational editor of Viking Press, shepherding to publication popular books on science.

All these things he did because he believed that science offered the children of immigrants a way to improve their meager circumstances, and it offered all Americans a means of advancing the ideals they held dearest. "The chief obstacles to science have always been fixed ideas and vested interests," he wrote in *Science and the Public Mind*: "Orthodoxies of all sorts, whether religious or political, whether moralistic or intellectualistic, are inimical to the spirit of inquiry. . . . An appreciation of the development of science as a great cooperative enterprise of mankind is likely to promote solidarity and to make each individual feel a sense of unity with his fellows. . . . Science is a means of broadening the sympathies and cultivating tolerance toward other groups, races, nationalities, tastes and philosophies."[81]

And what was true of science in general, according to Gruenberg, was especially true of evolutionary biology:

> The growth of the evolutionary point of view has carried a new implication regarding the significance of the individual. A new concept of species, as itself merely a convenient generalization, shifts attention back to the individual. The individual human being finds here a new dignity. He can articulate now excellent reasons for his resentment against being treated as a member of some inclusive group—whether it be peasant or teacher, voter or customer, or the more meaningless "public." He is discovering that he is after all, with the full approval of "science," what he always felt himself to be—namely, his own peculiar and unduplicated self. . . . There is always more in the individual than our categories acknowledge.[82]

This was the theme to which Gruenberg would return throughout his long career. In a elegiac essay entitled "Teaching Biology after the Wars," published in 1947, as the full extent of the Holocaust was still emerging, he wrote:

> Science is today the most dynamic agency for bringing the minds of men together. People everywhere are sundered and antagonized by their conflicting bigotries, their conceptions of virtue and worth, even by their criteria of beauty. But in science they can all reach common agreements. . . . Everybody [should] understand that what we call our present body of scientific principles includes contributions from all peoples; that it is both being used and continuously enlarged and refined by people of all nations, races, religious views; and that it is continuously being validated and reappraised by methods that are equally available to all people, regardless of race, nationality, or political or religious faith. . . . Science is . . . mankind's first systematic attempt to attain a clear and workable grasp on reality that has no secrets to hide and that excludes nobody.

At this moment in history, when delegations from nearly all nations are meeting, to find a basis for keeping the human race from destroying itself, the only groups that are able to agree are not the professional diplomats, not the religious leaders, not the orators and writers and scholars trained in history and the "humanities." It is the scientists alone who speak an international language and can agree on principles that affect our very existence. At the same time, scientists have generally, until recently, been either indifferent to what professional managers and leaders have been doing through politics or social criticism, or else quite lacking in strong convictions as to what kind of society science implies—nay, demands. Since the puffs of smoke blew away from Hiroshima and Nagasaki, some of us have come to suspect that the recommendations of scientists who are quite ignorant of the classic traditions could hardly be worse than the multiplicity of irreconcilable counsel to which we have been subjected.

The best we shall be able to do with our biology teaching may not end war—soon. But it may do much toward enlarging the areas of mutual understanding and mutual consideration. It might re-orient that self-assurance, not to say conceit, from which people generously try to be "tolerant" toward others. It might at least bring many more to heed Cromwell's pleas to his captains, when he said, "I beseech you, my brethren in the bowels of Christ, consider—you may be mistaken."[83]

Gruenberg, who devoted much of his career to teaching science in high schools or producing tools so that others could do so, saw public schools as sites for inculcating "strong convictions as to what kind of society science implies—nay, demands." Earlier, he and his wife, Sidonie Matsner Greunberg, had written in an article about child-rearing that "the free public school which Jefferson urged the Founding Fathers to include in the Constitution was conceived as necessary for a people that had freed itself—not from government but from a governing class—and was obliged thenceforth to govern itself. It was needed also to free the people from various provincialisms and superstitions and fears which they had accumulated through centuries of class-dominated stagnation. We should be freed of suspicions and hostilities toward strangers and toward new ideas." The Gruenbergs' essay concluded with a sort of homily: "Looking forward after this sketchy retrospect, the many issues seem to be reduced to one: either autocracy with a continuation of rule by authority or else a transfer through educational and democratic processes to government of common people by themselves, for their common benefit, through their own decisions and agents."[84]

For many American Jews in the first half of the twentieth century, immigrants and children of immigrants, the ideals Gruenberg hoped to spread by teaching science in public schools—democracy and meritocracy—had a compound appeal. There was the corrective they offered to the Century Club: ensuring that there would be realms of significance in which a Jew of Loeb's accomplishments would be exalted, not excluded, and leveling the field between Jews and America's Protestant elites. But there was also the corrective these ideals offered to the American Jewish Committee, an organization of self-appointed Jewish leaders who were alike in being wealthy, of German extraction, and unapologetically elitist. ("The monopolizing of leadership by a few," wrote Jacob Schiff, one of the founders of the American Jewish Committee, "should gradually be made to cease." But not soon.)

When the great waves of immigrants from eastern Europe arrived, they found a Jewish establishment that was committed to lending a hand in their settlement, but, at the same time, committed to keeping their scruffy co-religionists at arm's length. The condescension with which the "uptown" Jews regarded greenhorn "downtown" Jews was an engine of resentment throughout the first half of the twentieth century. It was an omnipresent fact of American Jewish politics and American Jewish identity. In 1930, the journal of the Jewish fraternity Zeta Beta Tau carried this breast-beating confession, demonstrating that uptown-downtown distinctions applied no less to the children of immigrants than to their heavy-tongued parents: "What have I done, with my wildly wielded weapon of the blackball, swung in my 17 year-old wisdom with the vehemence of a Cossack's thong? And for what? I did not like the cut of his coat or the part of his hair or his handclasp or his manner of speech. . . . I have blackballed a man who is of flesh and blood and bone and human sinew as much as I am, whose family is as decent and worthy as mine if not as refined and educated, for after all the step back to steerage is hardly more than two generations."[85]

Science offered a way out of this internal elitism as well. Consider the famous friendship of two of America's greatest physicists: Isidor Rabi and J. Robert Oppenheimer.[86] The men were young when they met; each had traveled to Germany, where the best physics was being done. They came from opposite backgrounds. Rabi had moved with his parents from Galicia to New York, where his father became a day laborer. In their slum apartment the language spoken was Yiddish and the religion practiced was rigidly Orthodox. Oppenheimer's father had come to the United States in 1888, as a teen, following an uncle who had already established a successful textile firm. Oppenheimer's mother, also of German-Jewish background, was an artist who taught at Hunter College. Oppenheimer was raised in a household rich in European culture and without religion. Rabi and Oppenheimer, then, were the very archetypes of the two sorts of American Jews who rarely interacted at work or play, who typically shared little in worldview and daily concourse. The distinction between uptown and downtown retained force in business, culture, society, education, and worship. But it did not in physics.[87] "Science has been a liberating art," Rabi wrote late in his life. "It has persuaded us to grow up and act our age in solving our own problems with our own means. What other liberating arts are there which will teach us to look at our problems objectively and solve them in a manner best suited to our needs and possibilities?"[88]

Oppenheimer shared this view. In 1953, a dozen years after he had taken management of the Manhattan Project, eight years after he'd witnessed its fiery denouement, and in the same year that he was stripped of his security clearance by the Atomic Energy Commission for his links with Communists decades earlier, Oppenheimer was invited by the BBC to deliver the prestigious Reith Lectures. The speech he broadcast, "Science and the Common Understanding," blended soaring appreciation for his vocation with melancholy. Men were capable of ignorance and bigotry, but they were also able to be rational and to progress. Man's long "journey of discovery," he said as he concluded his remarks, had led to a new "style of thought" and a new "understanding of community." With the Enlightenment, the journey had produced new scientific societies—"communities proud of their broad, nonsectarian, international membership, proud of their style and their wit, and with a wonderful sense of new freedom." Oppenheimer then quoted Bishop Thomas Sprat's description of the Royal Society: "Their Purpose is, in short, to make faithful Records of all the Works of Nature, or Art, which can come within their Reach. . . . It is to be noted that they [the members of this organization of savants] have freely admitted Men of different Religions, Countries, and Professions of Life. This they were oblig'd to do, or else they would come far short of the Largeness of their own Declarations. For they openly profess, not to lay the Foundation of an English, Scotch, Irish, Popish or Protestant Philosophy; but a Philosophy of Mankind."[89]

Oppenheimer added that "reading this today, we can hardly escape a haunting sense of its timeliness" because Sprat's "agreeable and noble" goal of using science as a cornerstone of a "philosophy of Mankind" had still not been achieved. "Our faith—our binding quiet faith," Oppenheimer concluded, ought be "that knowledge is good and good in itself" no matter who delivers it and what his religion.[90] The power of science, as Oppenheimer understood it,[91] owes in some measure to its laying good solid ground for a "quiet faith" that binds men and women whose more boisterous faith— their Judaism, Protestantism, Catholicism, Islam—too often rends asunder.[92]

Oppenheimer's conviction (and that of so many other American Jews) that science can and ought to be blind to the beliefs and backgrounds of scientists and the rest of us, has lately fallen into shabby disrespect. Popular faith in the good will of science and scientists has wavered. Hiroshima, Chernobyl, weaponized anthrax, climate change, and the many other products of scientific genius that now populate our nightmares have shaken the once-commonplace belief that science and technology are engines of human progress. Recent academic studies of the histories of science, medicine, and technology have tended to emphasize biases of race and gender, and focus on how science has served the monied and powerful at the expense of all the rest. And, in a very different way, religious partisans (advocates of teaching creation science in American public schools, for instance) have criticized science for its unfair "secular bias" that corrodes Christian piety. Science has recently become a battlefield with many fronts, attacked from the left and the right. In this fractious atmosphere, it is with some poignancy that one recalls how, in 1935, Gruenberg described science as "a means of broadening the sympathies and cultivating tolerance toward other groups, races, nationalities, tastes and philosophies." As he watched, brown-shirted thugs gaining power across the sea during that awful time, 1935, he insisted that the antidote to bigotry could be found in science. That so worldly and accomplished a man might have so naïve a thought at so dismal a time is poignant testimony to the faith shared by so many of his contemporary American Jews, that science might reshape the societies that embrace it into places where Jews, too, could thrive.

"Second Only to Communism"

Making Soviet Jews and Soviet Science

I N 1926, MOISEI GRAN, a Moscow physician and professor of medicine, together with other Jewish doctors of reputation, published the first number of a new scholarly journal they called *Problems of the Biology and Pathology of Jews*. The journal was a product of the Society for the Study of Social Biology and Psycho-Physics of the Jews, which had convened four years earlier.[1] Gran's own article in the inaugural edition recounted his research into how the "physical and biological appearance of Jews had changed as a result of moving from the city to the [Soviet] countryside" and described what scientists and social planners might learn from the changes. Resettling Jews and replacing their "non-productive" livelihoods with the heartier life of "the expansive green fields" of countryside collective farms, Gran explained, had changed the very nature of Jews.

This observation was not lost on Soviet politicians and planners. Nor did it go unnoticed even by many of the farmers alongside whom Jews had recently begun to work, rough-hewn sorts who had never held Jews in high esteem. Gran and his colleagues hoped, through their journal, to translate what the pols and the peasants had noticed into useful scientific fact.[2] Using modern science to elucidate "problems of the biology and pathology of Jews," Gran and his colleagues proposed, might achieve three salutary aims at once: advancing science, advancing the young Soviet Union, and advancing Jews.

These three aims were linked. For Soviet science, Jews were remarkable subjects of study: they were a great mass of poorly educated provincials who for generations had struggled to make livings through petty crafts, petty labor, or petty trade, and who after the revolution found themselves in new circumstances, some in the countryside, most in the large cities. What better way to gauge the impact of circumstances—of surroundings, education, language, food, work, and other factors—on cognition, perception, intelligence, health, and so many other things than to see how they affected the millions of Jews within Soviet borders? What better way to measure the value of the revolution itself than to document the distance it had taken Jews, from peasant pariahs (by and large) to productive citizens (by and large)? The editors of *Problems of the Biology and Pathology of Jews* had no doubt that their chosen subject was of importance in a great many realms. They wrote that "questions of demography, statistics, anthropology, questions of racial hygiene and eugenics, questions of physical constitution, inheritance, immunity, the most varied questions of social biology and pathology—all this can be studied most strikingly via the Jewish national organism."[3]

Problems of the Biology and Pathology of Jews was part of a larger Soviet scientific movement. Lev Semionovich Vygotsky, a scientist whose father had served as president of the Association for the Enlightenment of the Jews of Russia (*Obshchestvo dlia rasprostraneniia prosveshcheniia sredi Evreev Rossii*, abbreviated OPE), and who himself pioneered Soviet developmental psychology (what came to be known as "socio-historical psychology") also saw that the years of rapid change after the revolution offered a unique opportunity for science.[4] He sent his most famous student, another Jew named Alexander Luria, on an expedition to study peasants in Uzbekistan, with the aim of observing how Soviet education and collectivization changed the ways they reasoned.[5] "Perception and memory, imagination and thought, emotional experience and voluntary action [cannot] be considered natural functions of nervous tissue or simple properties of mental life," Luria came to conclude. "It [is] obvious that they have a highly complex structure and that this structure has its own socio-historical genesis."[6] Vygotsky, Luria, and their colleagues in what became known as the "Kharkov School" devoted themselves to documenting the plasticity of human character and cognition, developing an approach to psychology that has retained currency to this day.

All this mattered not just to scientists but also, perhaps especially, to the leaders of the young Soviet Union. The sort of research pursued by Gran, Vygotsky, and Luria (and many others as well)[7] was of practical importance for those seeking to make a modern nation with an advanced economy using the newest machines in factory and field, manned by the poorly educated sons of petty traders and peasants. "Civilizing" the masses of premodern sorts who made up the majority of the Soviet population after the revolution—and doing so quickly and permanently—was a task upon which the success of the entire revolution depended. It was the only way the new regime could fulfill the ambitious program that historian Francine Hirsch called "state-sponsored evolutionism": the goal was "to usher the entire population through the Marxist timeline of historical development, to transform feudal-era clans and tribes into nationalities, and nationalities into socialist-era nations—which, at some point in the future, would merge together under communism."[8] Many of the citizens of the new Soviet Union were of such primitive clans and tribes; understanding them, and discerning how they could best be transformed into productive moderns, was one of the greatest challenges facing the leaders of the revolution. Jews had a part to play in this, on both sides of the clipboard. Multitudes of Jews living near penury and illiteracy awaited transformation; growing numbers of Jews with advanced degrees worked to aid and document the change.

Problems of the Biology and Pathology of Jews illustrates in miniature why a great many Jews came to care a great deal about science in the first decades of the Soviet Union. Science offered a way for Jews to bootstrap themselves from the nether edge of society they occupied before the revolution to the center of what promised to be a modern, progressive, industrialized, urbanized Soviet Union. It offered a way for Jews to contribute to the success of their country, which was newly formed and newly willing to find a place for them. And science also offered a way to do well, to succeed and thrive in a new society and new economy that, for all the claims of equality, were proving from the start to produce winners and losers. Self-improvement, assimilation, citizenship, and success—science was a means to all of these for Jews after the Russian Revolution.

There was something deeper, too. Soviet science and Soviet Jews were formed at the same time, and from the very start, each was a part of the other. *Problems of the Biology and Pathology of Jews* offered an extreme example of a common phenomenon: Soviet Jews fashioning Soviet science in a way that helped fashion Soviet Jews and their place in Soviet society. Sociologists of science speak of the "co-production" of science and social institutions. In the first decades after the Russian Revolution, Soviet Jews and Soviet science took form together. Their stories are inseparable.

The Evolution of Soviet Jews and Soviet Science

The story of how the history of Soviet Jews and that of Soviet science became entwined begins long before the Russian Revolution. In the eyes of those who wrought it and those who fought it, the revolution was an upheaval that set everything on a new course. But matters were more complicated than this. The American abolitionist Wendell Phillips wrote that every revolution grows slowly, like an oak: "It comes out of the past. Its foundations are laid far back."[9] So it was for the new patterns of Jewish life and of scientific practice that evolved together in the first decades of the Soviet Union. In the final decades of the Russian Empire, the circumstances of Jews and those of science had already shifted in ways that set them on a path they would follow after the empire had collapsed.

RUSSIAN JEWS BEFORE THE REVOLUTION

At the beginning of the twentieth century more than five million Jews lived in the Russian Empire, accounting for 4 percent of the empire's total population and about 60 percent of Europe's Jews. Russian Jews were required to live in an area called the Pale of Settlement (including much of what are today Latvia, Lithuania, Ukraine, and Belarus); though some were exempted from this requirement, in the end about nine in ten Russian Jews made their homes there. Within the Pale they lived among themselves, half in small Jewish *shtetls* (or the country that surrounded them) and half in Jewish neighborhoods in bigger towns and cities. They spoke Yiddish and dressed differently than the Russian peasants and laborers among whom they lived. They stood out.

All this would soon change. Between 1897 and 1910, more than a million Jews moved from the country and small towns to cities. Russia at this time was undergoing rapid modernization and industrialization, and the same might be said for Russia's Jews. In the second half of the nineteenth century, the number of Russian youths who attended a rigorous secondary school, or *gymnasium*, grew more than six-fold, while the number of Jewish *gymnasium* students increased *fifty*-fold. "All the schools are filled with Jewish students from end to end," wrote Peretz Smolenskin, editor of the Hebrew-language Vienna journal *Ha-Shahar* in the early 1870s. "And to be honest, the Jews are always at the head of the class."[10] The change was even greater in universities. The overall number of university students in Russia increased by six times between 1840 and 1886, but the number of Jewish university students grew by more than a hundred times in the same years. One in three students at Odessa University in 1886 was a Jew (and over 40% of the students in the medical school), just as was one in three students at the Women's Medical Courses in Saint Petersburg.

The conditions that allowed this wave of Jewish enrollment in Russian universities had been set in place earlier. The first paragraph of the 1804 Statute on Jews—a document meant to set out how Jews were to be integrated in the Russian Empire—had declared that Jews were eligible for "all primary schools, gymnasia and universities," where they could study "medicine, surgery, physics, mathematics and other branches of knowledge" shoulder to shoulder with other subjects of the empire. Such a statute, an admiring journalist wrote at the time, would help "the state create useful citizens [through] moral upbringing."[11] For many reasons—including the suspicion and disinterest of Jews themselves, the reluctance of university sorts to accept Jews into their ranks, and the limited value a degree proved to hold for Jews—this liberal statute did not quickly lead Russian Jews to seek secular education in large numbers. Mordechai Aharon Ginzburg, a *maskil*, or scholarly devotee of the Jewish Enlightenment, the *Haskalah*, could still grouse in 1834 that among Jews, men of secular education "must conceal themselves like criminals." In 1840, of the 2,594 students enrolled in Russian universities, only 15 were Jews, and of these several converted to Christianity before receiving their degree.[12] It was not until later in the century, as the vitality of traditional, largely religious Jewish culture began to wane and the possibility arose that educated Jews might find their ways into Russian society, that Jews sought to enter the universities in great numbers and, via the universities, to enter liberal professions and the sciences.

In 1861, a Russian imperial decree accorded Jewish university graduates all the rights and privileges that Christian graduates received, including the rights to live where they chose and work in the profession for which they had trained. (A Minsk rabbi read the decree aloud in *Shul*, sermonizing to his congregation that "before us lies a bright, joyful future. Behind you— ignorance and death; before you—education and life. Choose!")[13] Following the decree, Jews flocked to universities. In 1865, Jews made up just over 3 percent of Russian university students. By 1876, the number had grown to 5 percent. Ten years later, it was almost 15 percent.

Indeed, by the last decades of the nineteenth century, the number of Jews in universities, and in the professions for which universities prepared them, had grown so rapidly that many of Russia's bureaucrats, pundits, and hoi polloi concluded that enough was enough. It did not go unnoticed, for instance, that a great many medals of honor and valor had gone to Jewish doctors for their service in the Russo-Turkish War of 1877–78, or that a great number of the physicians in cities like Odessa were Jews, or that even in a city like Saint Petersburg, 9 percent of physicians were Jews, as were 11 percent of the city's dentists and 20 percent of the pharmacists.[14] In the 1880s, quotas were instituted to limit the number of Jews in universities and in liberal professions. In 1882, the War Ministry limited to 5 percent the number of Jews attending the Military Medical Academy in Saint Petersburg. One year later, the Mining and Forestry Institutes of the Ministry of State Domains followed suit.[15] By 1886, so too had the Saint Petersburg

Institute of Communications Engineering, the Kharkov Technological Institute, and the Dorpat Veterinary Institute. The school of veterinary medicine in Kharkov banned Jews entirely.[16] In 1887, the tsar's Council of Ministers established a quota limiting Jewish enrollment to 10 percent in secondary schools and universities within the Pale, to 5 percent outside the Pale, and to 3 percent in Moscow and Saint Petersburg.

Overall, the number of Jews pursuing secular education declined, though never to the low levels stipulated by the quotas. In some places, it drastically declined; in Vilna, for example, Jews went from 58 percent of the student body in 1888 to 16 percent in 1900. As Chaim Weizmann, who grew up in Pinsk and went on to be a renowned chemist and, eventually, the first president of the state of Israel, remembered, the quotas "produced very curious, tragic-comic results. There were occasions when a rich Jew would hire ten non-Jewish candidates (at times rather oddly selected) to sit for the entrance examination at a local school, and thus make room for one Jewish pupil— needless to say, his own son or a protégé."[17] The quotas remained in force until the Russian Revolution, amounting to what one Jewish newspaper called a "silent, invisible pogrom" in higher education.[18]

This was a fraught turn of phrase, as the "silent, invisible pogrom" represented by the educational quotas had everything to do with the wave of clamorous and conspicuous pogroms Russian Jews had recently suffered. From 1881 to 1884, 200 Jewish neighborhoods and towns were sacked and pillaged under varied circumstances, a tsunami of violence ostensibly set into motion by rumors and newspaper reports that Jews had played a part in the assassination of Tsar Alexander II.[19] Although the attacks were condemned by the Russian government and, in some circumstances at least, the Jews were protected by government troops and by police, the popular sentiment behind the violence led in part to the quotas and sanctions meant to ensure that Jews remained pariahs. Another flood of attacks began in Kishinev the day before Easter 1903. The *New York Times* reported that the frenzy was premeditated, beginning on the eve of the holiday with

> a well laid-out plan for the general massacre of Jews on the day following the Orthodox Easter. The mob was led by priests, and the general cry, "Kill the Jews," was taken up all over the city. The Jews were taken wholly unaware and were slaughtered like sheep. The dead number 120 and the injured about 500. The scenes of horror attending this massacre are beyond description. Babies were literally torn to pieces by the frenzied and bloodthirsty mob. The local police made no attempt to check the reign of terror. At sunset the streets were piled with corpses and wounded.

Those who could make their escape fled in terror, and the city is now practically deserted of Jews.[20]

For more than three years, pogroms of this sort were reenacted in hundreds of towns and villages, killing thousands, wounding many more, and destroying incalculably much of the property that the generally hardscrabble Jews had accumulated.

It is tempting to see the spasms of violence against Jews in the last decades of the Russian Empire as a continuation of the long history of anti-Semitism in eastern Europe. It would not be wrong to do so. But for all the anti-Jewish violence that had existed in Russia in the past, the violence of the late nineteenth and early twentieth centuries also represented something new, refracting uneasy changes in the empire itself. Among these changes were halting and ambivalent efforts to better integrate Jews into Russian society. Also playing into these unsettled times was the deterioration of imperial control and imperial rule over the vast holdings of Russia. The pogroms in the final decades of the empire also reflected anxious concern that Russia's army, its agriculture, its factories, its universities, its cities, and all the rest were falling ever further behind those of Germany and the rest of Europe. Looking back, it is possible to see that it was when Russians were most insecure that Jews suffered the most: after Alexander II's assassination, for example, and after the hapless Russo-Japanese War in 1904 and 1905.[21]

Whatever the reason, during the quarter century bracketing the two waves of pogroms, more than a million Jews fled Russia, some to Palestine, some to Britain and most to the United States. Many of those who did not leave Russia tried at least to flee the Jewish towns and villages of the Pale of Settlement and make their way to the cities; here, there would be greater anonymity, which might mean greater safety as well as expanded opportunity.

In the final years of the empire, then, Russian Jews were under tremendous pressures from within and without. They were a population that had changed almost beyond recognition over the course of a generation and that was certain to continue to change rapidly in the future. The last of the tsars had launched a policy that Benjamin Nathans has called "selective integration," fitfully allowing Jews in great numbers to enter universities, or to hold down jobs in civil service, or to advance—to a point—within the army. As a result of this policy, Nathans writes, "there were in effect two Russian Jewries: the mass of legally and culturally segregated Jews confined to the Pale and a small but growing number in and beyond the Pale whose integration into the upper reaches of the surrounding society (though certainly not into the ruling elites) was proceeding far more rapidly than anyone had expected."[22]

The new class of educated, cosmopolitan Jews, living in heartland cities, represented a new ideal for Russia's Jews. But despite their success relative to those they left behind in the Pale, they were hardly a satisfied lot. They could not overlook the assaults that Jews suffered or avoid contemplating what those attacks said about the place of Jews in Russia.[23] Violence aside, the pace of "emancipation" for Jews in Russia remained reliably slower than it was elsewhere in Europe, and Jews looking westward expressed frustration at lagging behind. This was not an idle complaint. The faltering and inconsistent delegation of rights to Jews could produce circumstances of comic irony. In 1906, Shmaryahu Levin was voted into the Duma. But although the law allowed Jews to seek election, it did not allow them to enter the city in which the Duma convened.[24] For another thing, by the final gasping years of the empire, a Jew might justly conclude that matters were growing bleaker with every passing year. Because of quotas, universities and the advancement they offered were less accessible than in prior decades. Employment in the civil service grew harder, not easier, to attain. Permission to quit the Pale for cities became harder to acquire.

Some whose fondest wish had earlier been Russification came to conclude that more good would come of trying to modernize Judaism, and especially Yiddish culture, without abandoning it.[25] Many other Jews turned to revolutionary movements. Abraham Cahan wrote that it was pogroms that caused Jews to embrace revolution: "The white terror of the knout, the prison cell and the gallows gives birth to the red terror of the pistol and the dynamite bomb."[26] The Algemeyner Yidisher Arbeter Bund in Lite, Poyln un Rusland (General Jewish Workers' Union in Lithuania, Poland and Russia), or Bund, as it came to be known, was founded in Vilnius in 1897 with the seemingly paradoxical aim of uniting all the empire's Jews under the red flag of a single socialist party. There, they would gain the class consciousness needed to relinquish their narrow identification as Jews once and for all. The formation of the Bund had become possible because of the changes in the demography of Russia's Jews in the late nineteenth century. More Jews worked in industry than ever before, more lived in cities, and more lived far from the towns and neighborhoods in which they were born. If the appeal of socialism is most potent to workers whose ties to the traditional culture in which they were born have been stretched or severed and who have little stake in the factories that overwork them, it is no surprise that it found a powerful foothold among Jews in the waning years of the empire. In general, the balkanization, dislocation, secularization, politicization, radicalization, and urbanization experienced by many of Russia's Jews at this time left them in an excited and volatile state, free radicals waiting for something to bind to. Looking back on this time from the vantage point of 1923, poet Shmuel Halkin wrote:

> Russia! If my faith in you were any less great
> I might have said something different
> I might have complained: You have led us astray,
> And seduced us young wandering gypsies.[27]

Benjamin Nathans' conclusion about this period is one of dry understatement: "As it entered the twentieth century, Russian-Jewish society was thus the site of extraordinary fractiousness . . . as well as extraordinary ferment."[28]

RUSSIAN SCIENCE IN THE LAST DECADES OF THE EMPIRE

If Jewish life and livelihood changed in the final decades of the empire, so too, in its way, did Russian science. In 1911, the great plant physiologist and popularizer of Darwin, Kliment Arkadievich Timiryazev, could still lament that primitive Russia kept "all its science concentrated in universities." This was in contrast to "the entire civilized world," where science thrived above all in independent research centers (like the Kaiser Wilhelm Institutes) on the one hand, and in industrial and commercial laboratories on the other.[29] Timiryazev's complaint reflected odd anomalies in the way science developed during the final decades of the empire. It was true that Russian science thrived best at government-supported universities and within myriad learned societies, and that the more rarified institutions of pure research that flourished in Germany, Britain, and France and were beginning to take root in the United States (spurred by the Carnegie Foundation, above all) did not evolve in Imperial Russia. And it was true that Russian science failed to find a foothold in Russian factories, which relied heavily on chemicals and dyes trucked in from Germany.

There were many reasons why this was so, some practical, some political, and some philosophical. Practically, the sorts of individuals and institutions who in other places were spending fortunes to promote science were fewer in Russia. Perhaps they were also less intrigued and inspired by the wonders of science than their counterparts in western Europe and North America. Timiryazev had already complained in 1884 that science had not implicated itself into civil society in Russia with the same force it had elsewhere.[30] Politically, the late imperial government was reticent to allow alternative centers of education and research to arise, and was quick to see in the professoriat a source of threat. This fear was not unwarranted because although professors had traditionally been a stodgy and conservative bunch, scientists and professors could now be found in outsized numbers among reformers and even revolutionaries. Indeed, Timiryazev and other prominent scientists like Vladimir I. Vernadsky argued publicly and *ex cathedra* that science thrived where democracy thrived.[31] "There [are] decent people in the professoriate," Tsar Nicholas II said in 1904, "but very few."[32]

Philosophically, Russian philosophers of nature and scientists had long embraced an ideal of "pure" science, one that viewed the material applications and benefits that might accrue from science as disreputable by-products of a noble endeavor. As historian Alexei Kojevnikov wrote, "In fin-de-siècle Russia the ideology of pure science was taken much more seriously and literally" than elsewhere in Europe.[33] This preference was a long-standing tradition in Russian science and natural philosophy (and fits well with the idealistic bent of much Russian philosophy and literature). The great zoologist and evolutionary theorist Peter Kropotkin chided his younger colleagues in 1885, writing: "The day when you are imbued with wide, deep, humane and profoundly scientific truth, that day you will lose your taste for pure science. . . . You will place your information and your devotion at the service of the oppressed."[34]

It is a perennial parlor game among historians of science to hypothesize about the degree to which science in various times and places had an identifiable "national character." Is there anything French about French science, or British about British science, or German about German science? Many believe that sometimes it is possible to answer these questions in the affirmative but that such national character as exists is at best a weak force buffeted in a field of many other forces. One can easily imagine that a culture that idolized Dostoyevsky and Tolstoy was especially hospitable to science of a more theoretical bent. ("I don't think it is dishonorable for a Russian professor of chemistry to work in the applied direction," Vladimir Markovnikov of Moscow University wrote in 1901, suggesting in a doth-protest-too-much sort of way that it *was* dishonorable, at least by some people's lights.)[35] But the notion that Russian scientists preferred "pure" over applied science reflects a Russian bent for idealistic abstraction may be overly fancy. The theoretical bent of Russian science was surely reinforced by the more earthly fact that in fin-de-siècle Russia there was not a lot of practical, applied, and impure science with which a young aspiring professional could occupy himself and pay his rent. In fact, the practical, political, and philosophical reasons for science remaining in the province of university halls and rarefied honorific professional associations each seemed to reinforce one another.

That is, until they stopped. By the first decade of the twentieth century, there were many signs that the established ways of organizing science were doddering. In January 1905, 125 professors, 201 junior faculty members, and 16 members of the Russian Academy of Sciences published a manifesto they called the "Declaration of the 342," which read in part:

> A stream of government decisions and regulations has reduced professors and other instructors at the institutions of higher education to the level of bureaucrats, blindly executing the orders of higher government authorities. The scientific and moral standards of the teaching profession have been lowered. The prestige of educators has dwindled so much that the very existence of the institutions of higher education is threatened. Our school administration is a social and governmental disgrace. It undermines the authority of science, hinders the growth of scientific thought and prevents our people from fully realizing their intellectual potentialities.
> . . . Academic freedom is incompatible with the existing system of government in Russia. The present situation cannot be remedied by partial reforms but only by a fundamental transformation of the existing system.
> . . . Only a full guarantee of personal and social liberties will assure academic freedom—the essential condition for true education.[36]

This declaration must be seen in the broader context of the Revolution of 1905 and the efforts to democratize all aspects of life under the tsars. By this time, the universities were fully identified with calls for social and political reform, which for a time seemed to be right at hand. The Duma was established, and though it had little authority, it signaled progress from autocracy to democracy. In October 1905, Count I. I. Tolstoi was appointed minister of education, and he immediately set out to put the universities on a new footing, describing his goals in an internal memo halting the "Russification" policies of the ministry, abolishing restrictions on Jewish students, and granting greater autonomy and self-government to university faculty.[37] These policies were controversial at the time and were only partially put into practice. In any instance, Stolypin's Coup in 1907 reversed most of the changes that had resulted from the Revolution of 1905. But the upheaval on campuses and beyond never cooled to less than a simmering, low boil.

Amid the continuing ferment, Russian scientists grew more confident and full-throated in their demands for greater budgets and greater autonomy. By 1911, Timiryazev could reasonably expect the assent of his colleagues when he wrote that "the success of science (and technology) is impossible without emancipating the modern scientist from his obligations as a teacher."[38] Indeed, envy of the newly built Kaiser Wilhelm Gesellschaft in Berlin was at the heart of an emerging consensus among Russian scientists that their government too must establish and support independent scientific research institutes that made no demands on their faculty save that they advance Russian science. The very nature of imperial science—the small-scale university study of pure science—had come to seem quaint and inadequate, even in the eyes of scientists who had managed to attain campus sinecures.

The conclusion that Russian science simply had to change became all but undeniable with the outbreak of the Great War. The war showed with terrifying clarity how much Russian science and Russian manufacturing depended upon German pure and applied sciences. Industries that relied on scientifically advanced new technologies discovered with a start that more than half of their chemicals and machines were developed and produced in Germany. When the border with Germany closed in August 1914, a great deal of Russian industry imploded like a defunct Vegas hotel. Foreign investors owned most of Russia's civilian industries, and they had never bothered to establish research and development functions in Russia. Even state-owned arms and munitions factories would buy and copy innovations from elsewhere rather than pay scientists and engineers to develop technologies at home.[39] Writing in 1915, at the height of the crisis, the geochemist Vladimir Vernadsky wrote: "Russian society has suddenly realized its economic dependence on Germany, which is intolerable for a healthy country and for an alive strong nation. . . . [This dependence] has developed into an exploitation of one country by the other. . . . One of the consequences—and also one of the causes—of Russia's economic dependence on Germany is the extraordinary insufficiency of our knowledge about the natural productive forces with which Nature and History has granted Russia."[40]

This dependency was in evidence across the board. The frustrations of the editors of the journal *Priroda* were on full display when they wrote in 1915 that "until now our country has made no serious effort to produce its own scientific and educational instruments and to free itself from the stranglehold placed on it by Germany."[41] The war drove home the inadequacy of Russian science and spurred changes on the fly that would have a lasting impact on Russian (and, soon, Soviet) science. Russian chemical companies went from employing 33,000 workers 1913 to 117,000 workers in 1917. Writing in 1914, Moscow chemistry professor Aleksei Chichibabin insisted that the rapidly growing Russian chemical industry, "from the very beginning, must find its basis in Russian science . . . and take care of the establishment of most favorable conditions for the quickest and widest development of Russian chemical science."[42]

And Russian science grew at an unprecedented clip, as indeed it needed to do. If in most fields Russia was mostly a provincial client of European science, especially that of Germany, with the war, contact, communication, and collaboration across the border suddenly stopped. If Russian research before the war was written in German or English and printed in foreign journals, during the war new Russian-language periodicals were established. So too were new national scientific societies. The internationalist tendencies of Russian scientists before the war were quickly replaced by new nationalist ones. Old preferences for pure science dissolved in the face of the insatiable needs of the military and of a society now at war. Vernadsky himself proposed that the Imperial Academy of Sciences retreat from its commitment to pure science and establish what he called a "Commission for the Study of Natural Productive Forces of Russia" (KEPS). His notion was that the commission would do in an organized fashion what the military had begun haphazardly through its "mobilization of various engineers who work on the basis of exact sciences, physicians, bacteriologists, and . . . chemists": put science to work in the service of the Russian people. "After the war of 1914–1915," Vernadsky wrote with an eye to the future, "we will have to make known and accountable the natural productive forces of our country, i.e., first of all to find means for broad scientific investigations of Russia's nature and for the establishment of a network of well equipped research laboratories, museums and institutions. . . . This is no less necessary than the need for an improvement in the conditions of our civil and political life, which is so acutely perceived by the entire country."[43]

Vernadsky could already see that the science of Russia's future would be different from the science of Russia's past: it would be carried out by different people, for different purposes, in different sorts of laboratories, with different results.

He was right on every count. At the same time, it is now easy to see that many of the changes that in time came to be associated with the Russian Revolution were well under way before the Romanovs were placed under house arrest at the Alexander Palace or before the ten days in October that shook the world. This was true of almost every aspect of Russian society, including the revolution's effect on the Jews, on science, and on the Jewish involvement in science.

Jews and Science after the Revolution

Still, it was not until after the Revolution that many changes that had long been in progress became embedded in Russian, now Soviet, society. One change was in the official status of Jews. Just after the October Revolution, Fred Haggard, who knew the country well after his posting there as senior American secretary for War Prisoners Aid (WPA), observed in an essay called "The New Spirit in Russia" that

> the treatment of the Jews was another illustration of the working of the former government. Their treatment marked the very climax of autocratic hate and senselessness, and it is the very irony of fate that now brings to the front men of that race to vex the souls of those who made pogroms and nameless horrors possible. The sad part about all this was that it was not simply the government, but the church, that instigated and carried through these cruelties and thus denied the principles and undermined the basis upon which it was supposed to be founded. Religious liberty had no place in the old Russian scheme of government. Today, religious liberty is an absolute fact in its fullest meaning. We have no more perfect religious liberty in America than there is in Russia at this moment.[44]

It is hard to resist the tendency to view early Soviet society by the light of what we know came to pass with the year. Anti-Semitism never disappeared in Russia, and not many years passed before it was cannily exploited by Stalin. Still, the revolution offered real hope and, more important, genuine advancement to Russia's Jews. Lenin said:

> The Tsarist police, in alliance with the landowners and the capitalists, organized pogroms against the Jews. The landowners and capitalists tried to divert the hatred of the workers and peasants who were tortured by want against the Jews. . . . Only the most ignorant and downtrodden people can believe the lies and slander that are spread about the Jews. . . . They are our brothers, who, like us, are oppressed by capital; they are our comrades in the struggle for socialism. . . . Shame on accursed tsarism which tortured and persecuted the Jews. Shame on those who foment hatred towards the Jews, who foment hatred towards other nations.[45]

Officially, Soviet law outlawed religious and ethnic discrimination, and while such discrimination never fully disappeared, the case can be made that Jews found fewer obstacles to obtaining education and employment in the USSR in its first decades than probably anywhere else on earth. A symbolic token of this change is the fact that of Lenin's four closest allies in the revolution, three—Trotsky, Zinoviev and Kamenev—were Jews by heritage.

But there was more than symbolism in play. The revolution had crushed the old elites of tsarist Russia, and discrimination against the children of the prerevolutionary monied and powerful continued long after the revolution was won. This created opportunities for previously downtrodden and im-

poverished minorities, opportunities exploited more vigorously by Jews than anyone else. For one thing, Jews remained by far the most literate group in the USSR (85% of Jews could read in 1926 and 94.3% in 1939, compared with 58% and 83.4% for non-Jewish Russians). By 1939, a Soviet Jew was more than three times as likely to finish secondary school as the general population. Seventeen percent of university students in Moscow were Jews, as were 19 percent in Leningrad, 24.6 percent in Kharkov, and 35.6 percent in Kiev. Jews were ten times as likely to complete university studies as the general population, and one of every three college-age Soviet Jews was studying at a university.[46] These trends ran deep and would prove durable: by 1989, at the very end of the Soviet epoch, 64 percent of Jews had a university education, compared with 15 percent of Russians.

Most of these Jews went into scientific and technical fields. A 1937 survey found that of the 82,300 Jews attending Soviet universities, 30,900 (37.5%) were in technical faculties.[47] It was estimated in the early 1950s that 11 percent of all Soviet scientists were Jews, though Jews made up only 1.5 percent of the population.[48] In 1959, Jews were more than thirteen times as likely to be "scientific workers" as Russians were. The top five Jewish occupations were engineering, medicine, work as "scientific personnel," teaching (often in math and science), and chief production and technical management; the first three of these occupations accounted for more than a quarter—28 percent—of all Jewish employment. It was after the Soviet Union disbanded and almost a million Jews immigrated to Israel that the remarkable Jewish embrace of sciences in the USSR became most conspicuous. Over 60 percent of those immigrants who were employed prior to their immigration were engineers, scientists, physicians, nurses, technical workers, and other professionals (this last category comprising mostly teachers, most of whom taught scientific disciplines).[49]

These statistics alone are enough to show that Jews were prominent in Soviet science. But there are two sorts of prominence. One sort, as we have seen, is numerical. A great many scientists, doctors, and engineers in the Soviet Union were Jewish. The second sort concerns eminence. Here, too, statistics help paint the picture. For instance, four of the seven Soviet Nobelists in physics were Jews (and more than 30% of all Soviet prizewinners up to 1975).[50] But to take full measure of the eminence of Jews in Soviet science, one must look beyond the statistics to the individuals who molded and guided Soviet science in its first decades, many of whom were Jews.

The eminence of Jews in early Soviet science is rarely remembered today, but it was a matter of some interest at the time. In his waning days, the great Hungarian Jewish physicist émigré to the United States, Edward Teller, reminisced about a road trip he and his wife took with George Gamow, the equally renowned Soviet physicist and émigré to America, and his wife:

> We arrived in Florida and the closer we got to Miami the more annoyed Gamow seemed to get. I didn't know why. I didn't understand. Then his pretty wife Rho explained to me. "You know, you may not have noticed, Gamow is anti-semitic; there are too many Jews here." Gamov anti-semitic! His best friend in Russia was the Jewish Lev Landau! . . . That was one of the few occasions when he and I talked about politics. What he meant was that he was terribly, terribly unhappy about the Soviets, and Stalin, about the Communist government and he saw a connection, not a close one, . . . between Jews and Communism.[51]

That a Soviet physicist of Gamov's stature linked the Soviet regime with Russian Jews is not as odd as it may seem from today's remove. Teller observed that Gamov's friend and colleague was the Jewish physicist Lev Landau. In the years when Soviet physics first established itself—the years when Gamow came of age as a physicist—a majority of the most important physicists were Jewish. Landau was the rule, not the exception, as Gamow well knew. Among Landau's mentors, colleagues, and students (and Gamow's as well) were a remarkable number of Jews who established and ran some of the most important science institutes of their day, making Soviet science—above all, physics—into the international powerhouse that it became: Abram Fyodorovich Ioffe (1880–1960), Yakov Il'ich Frenkel (1894–1952), Matvei Petrovich Bronstein (1906–38), Leonid Isaakovich Mandelshtam (1879–1944), Igor Yevgenyevich Tamm (1895–1971), Semion Petrovich Shubin (1908–38), Yuli Khariton (1904–96), Bentsion Moiseevich Vul (1903–85), Fedor Galperin (1903–?), Yuly Borisovich Rumer (1901–85), Semen Alexandrovich Altshuler (1911–83), and many more.

Although any one of these men might contend for the designation, it is Abram Ioffe who is most often called the "father of Soviet physics," and for good reason.[52] Ioffe won adulation in his lifetime and after his death, receiving the Stalin Prize (in 1942), the Hero of Socialist Labor Prize (in 1955), and the Lenin Prize (posthumously, in 1960); he founded some of the most important research institutes in the Soviet Union, mentored generations of the country's best physicists, and carried out pioneering research, to international acclaim, over the course of half a century. Yet nothing in his background suggested that such extravagant success was possible. In the Ukraine of 1880, where Ioffe was born to a Jewish petty merchant of modest means, there were no passable trails which led to the elite of intellectual society. Ioffe received a generally inferior education; and when the time came, he enrolled, not in a classical gymnasium, but in a technical high school, a circumstance that all but barred him from Russian universities. Instead, he attended the Saint Petersburg Technological Institute. When he graduated in 1902, his professors exhorted him to continue his studies in Germany (like many other Russian Jews who sought advanced education in the sciences), and Ioffe procured a practicum in Wilhelm Roentgen's renowned X-ray laboratory at Munich University.[53] Ioffe remained there for most of four years, earning a prestigious doctorate for his discovery of a photoeffect (the "elastic aftereffect") in X-rayed quartz, a précis of which was published in the *Annalen der Physik*.[54] But this too was not enough to secure him a professorship upon his return to Russia, where advanced degrees were recognized only for those scholars who held a high school diploma from a classical gymnasium, and where Ioffe's Jewish background made him harder to hire. Ioffe became a laboratory assistant at the recently chartered Saint Petersburg Polytechnical Institute.[55]

New institutions often offer opportunities to talented outsiders that more established ones, bound in tradition and concerned for reputation, do not. It was not long before Ioffe was invited to teach at Saint Petersburg Polytechnical, and he soon began to publish on a variety of topics, such as the photon theory of radiation. Indeed, he continued to work with Roentgen, traveling frequently to Munich, and building an international reputation that eclipsed his more constrained local one. (His ties with Roentgen were severed only with the outbreak of the Great War.)

At the same time, Ioffe began to cultivate an intellectual life of weight and substance at home. Soon after his return from Germany, Ioffe was joined in Saint Petersburg by a young Austrian Jewish physicist, Paul Ehrenfest, who had followed his Russian wife, mathematician Tatiana Afanas'ev, back to her home. A foreigner and a Jew, Ehrenfest could wrangle only a temporary, part-time job with no hope of tenure, but still, his influence on Russian physicists, Ioffe first among them, was enormous. Ehrenfest and Ioffe met frequently to discuss developments in physics (during arguably the most exciting years in the discipline's history). The two men were outspoken members of the Russian Physical and Chemical Society and leaders of the young Turks advancing the newest theories. Together, they were the motive force of what has been called the Petersburg Physics Seminar, a group of young scholars who met to discuss cutting-edge physics as well as to lay the groundwork for new and improved institutions through which physics could be researched, studied, and taught. They aimed to build a central scientific institute in which to pursue research, and to devise new programs in which the best up-and-coming Russians could learn physics without travel abroad. In 1912, however, Ehrenfest finally left Russia and was soon thereafter selected by H. A. Lorentz to be his successor at Leyden University.

Ioffe stayed, teaching and at the same time pursuing further advanced degrees at Saint Petersburg University, in the hope that these would allow him to assume a professorship. The great Soviet inventor Leon Theremin (or Lev Sergeyevich Termen, as he was known in Russia) recounted how in 1913, when he was just seventeen, he was invited by a cousin who knew of his interest in physics to sit in on the master's thesis defense of Ioffe, who was forging a reputation as a rising star. Theremin and his cousin had to jostle to find seats in the crowded room. After Ioffe's jury passed him with superlatives, Theremin's cousin introduced the boy to the young physicist: "I thought how good it would be if I could work under his guidance, had I already grown up."[56] (It was a wish that would later be granted.) Soon thereafter, Ioffe was secure enough in his future in Petersburg that he turned down a professorship at Kharkov University.[57] By 1915, Ioffe had completed a doctoral dissertation and was promoted to full professor at the Polytechnical Institute. In 1916, Ioffe and his Saint Petersburg colleagues began to publish pathbreaking articles on crystallography, based on their X-ray work, in what amounted to "the first works of his emerging research school."[58]

With the October Revolution, nothing changed for Ioffe's career—and then again, everything did. Just over half a year after the revolution, Ioffe was selected to direct the Physico-Technical Division of the new State Roentgenological and Radiological Institute. The notion of such an institute was itself at once new and old. In 1910, a Jewish physician and roentgenologist, Mikhail Isaakovich Nemenov (1880–1959), had proposed establishing such a center. Nemerov, who came from Vitebsk and had completed a degree at the Petersburg Medical Institute,[59] was convinced, like many of his contemporaries, that independent institutes of the sort he labored to establish were the only way to tow Russian science and technology into the twentieth century. In 1917, Ioffe joined Nemenov's efforts, and almost immediately after the revolution the men persuaded Anatoly Lunacharskii, head of the new People's Commissariat of Education, to make their vision a reality. And so it was that, as one historian observed, "by the time the establishment of the State Roentgenological and Radiological Institute (part of which would later become Ioffe's Physico-Technical Institute) was officially announced on May 6, 1918, it had long been actively functioning. Ioffe was already carrying out his classical investigations there on the structure of mechanically deformed crystals."[60]

Thus, the new institute's work, while it was in some ways a continuation both of Ioffe's vision and research, was also a new beginning—for Ioffe and for Soviet science. Even as the Bolsheviks were consolidating their control over the USSR, Ioffe and colleagues who shared his aims were hastily trying to consolidate their own control over their new country's physics. They established the Russian Association of Physicists (RAF, by the Russian acronym), and Ioffe was selected its president. They petitioned the People's Commissariat for Education, commonly called Narkompros, for authority to establish a new research center, The State Roentgenological and Radiological Institute (GRRI); and in May 1919 Zorakh Grinberg, the Jewish representative of the new commissariat, granted permission to incorporate. The institute had four departments: Nemerov headed the medico-biological sector, and Ioffe took charge of the physico-technical sector. Almost immediately, Ioffe's "sector" began to operate as an independent institute itself, and an incomparably important one at that.

That the People's Commissariat for Education chose to support scientists who first found work in the tsar's universities is not a fact one can take for granted. Other intellectuals who had the bad luck of establishing themselves in the tsar's Russia—writers, philosophers, historians, poets, and more—were viewed by the revolutionary guard with suspicion. The Workers' Opposition, the Proletarian Culture Organization (*Proletkultists*), and the many other groups that burst to life, animated by revolutionary fervor to replace the debauched old with the pristine new, demanded class war against "bourgeois remnants," of whom those who frequented university lecture halls offered an exemplar.[61] In 1922, Lenin ordered the exile of two shiploads of such intellectuals in what was known as the "voyage of the philosophy steamer." Among the exiled intellectuals was Semyon Frank, a famous philosopher who had converted from Judaism to Christianity and whose nephew, Ilya Frank, achieved even greater fame as a Soviet physicist, eventually winning the Nobel Prize in 1958.[62] Scientists of Ioffe's generation benefited from a double standard that saw in physicists, chemists, biologists, and the rest an inherent simpatico to the revolution that philosophers and poets did not have. Lenin himself made a double argument for treating scientists as valued allies instead of suspect traitors. In practical terms, the revolution and the Soviet Union needed what scientists alone could offer—the means to quickly advance the country's industry, agriculture, and development. In terms of ideology, Lenin argued, scientists were by their nature materialists because modern science and technology militate to materialism. Positivism, Lenin believed, was a gateway to Marxism. The very vocation of scientists would draw them in time to socialism.[63]

Ioffe's career, in any case, thrived under Communism, as did the fortunes of the Leningrad Physico-Technical Institute. There are many measures of the fecundity of Ioffe and his center. In its first twenty years, ten new institutes sprouted from the original. Many dozens of academicians and corresponding members of the Soviet Academy of Sciences worked or studied at the institute.[64] Many of the Soviet Union's greatest physicists trained there, worked there, or both, including the Nobel laureates Nikolai Semenov, Ilya Frank, Igor Tamm, and Lev Landau. That Ioffe's institute produced so many laureates says something about the place. That three of the four men—Frank, Tamm and Landau—were, like Ioffe himself, born to Jewish parents says something about the penetration of Jews into Ioffe's institute. Indeed, of the institute's original staff of eight, at least three came from Jewish backgrounds.[65] Consider the moment of the founding of the State Roentgenological and Radiological Institute, in which a revolutionary functionary of Jewish heritage charters a research center at the urging and under the direction of a physician and physicist of Jewish heritage. Fifty years earlier, even five years earlier, this would have been unthinkable. In no time after the revolution, Jews found a place in the heartland of Soviet physics and, more broadly, Soviet science.

Of course, this was not a phenomenon limited to Leningrad. In the winter of 1934/35, British journalist James Gerald Crowther made his famous survey of the Soviet science establishment and found that a great many of the new research institutes were directed by Jews. Nemerov, I have already noted, directed the Roentgenological and Radiological Institute alongside Ioffe.[66] The Physico-Technical Institute that was established in Kharkov was run by Lev Landau.[67] At the head of the parallel institute in Dnepropetrovsk stood a student of Ioffe by the name of B. N. Finkelstein.[68] Boris Hessen directed the Physical Institute of the University, Moscow.[69] Yakov Dorfmann ran the Physico-Technical Institute of the Urals, Sverdlovsk. David Talmud headed the Laboratory for Surface Chemistry in Leningrad's Institute for Chemical Physics.[70] The Maxim Gorky Medico-Biological Research Institute in Moscow was run by Solomon Levit.[71] Ioffe's friend and former student, Yakov Frenkel, became chairman of theoretical physics at Ioffe's alma mater, the Leningrad Polytechnic Institute.[72] V. F. Kagan became the head of the Scientific Division of the State Publishing House.[73] Lev Landau directed the department of theoretical physics at the Kharkov Polytechnical Institute and then went on to direct the Theoretical Division of the Institute for Physical problems in Moscow.[74] Semen Aleksandrovich Altshuler became head of theoretic physics at Kazan University.[75] Yuli Khariton directed Arzamas-16, the center for atomic bomb research.[76]

It would be easy to make too much of these facts. Russian science grew quickly, and there were many institutes and departments to head. Measured against the sum total, the impression that Jews *dominated* Soviet physics quickly dissipates. They did not. But it is possible to make too little of these facts as well. For they do show that the place of Jews in science, in intellectual life, and in society more broadly changed immeasurably in the generation that divided the pogroms of the 1880s from the period I have been describing, the first decade or so after the revolution. The path from the Pale to the Party and the professoriat had been so short that grandparents born of a caste banned from cities, rejected from universities, and ineligible to work in the civil service survived to see their grandchildren heading some of the most revered institutions in the great metropolises of the country. In the Russia of 1880, no one could have predicted, or even imagined, the preeminence to which Jews would rise in the Soviet Union of 1930.

Toward an Explanation of the Surprising Success of Jews in Soviet Science

There were many reasons that Jews ascended so quickly in science, and to such heights. The first, and arguably the most important, was chance. After the revolution, the Soviet state was desperate for scientists and technicians and eager to build new laboratories and institutes in which they could train and work. This was, to a great degree, an extension of the pained realization to which earlier Russian leaders had been brought by the Great War, the realization that the state's very survival depended on building the infrastructure of engineering, science, and manufacturing that the country lacked. What's more, the Bolsheviks understood that the sustainability of their revolution depended on their ability to feed and clothe the multitudes they now ruled and that they could do this only if they managed to build modern factories and farms, which, in turn, they could do only if they had talented scientists and engineers working in labs stocked with up-to-date equipment, aided by talented students of intellect and energy. They understood that for the Soviet experiment to work, as many scientists as possible needed to be pressed into its service as soon as possible. Scientists who had established themselves before the revolution needed to be supported; and existing labs, research centers, and university programs needed to be supported. But these things were not enough. New scientists, new scientific workers, and new workplaces were also needed.

Jews, especially the multitude of young Jews who made their way to Russian cities from the Pale of Settlement, were especially fit to become scientists and engineers and lab technicians and doctors. They were the most literate of the ethnic groups in the Soviet Union. They were, as historian Yuri Slezkine observed, "the only members of the literate classes not compromised by service to the tsarist state (since it had been forbidden them)."[77] This mattered, because they brought with them no great loyalty to, little affection toward, and scant nostalgia for an imperial regime under which they had suffered, often extravagantly. They were harder to suspect of harboring antirevolutionary sentiments than other "bourgeois experts" who might have something to gain by a return to the status quo ante. Their alienation from the past was an asset for the present and future, as Lenin himself observed: "The fact that there were so many Jewish intelligentsia members in the Russian cities was of great importance to the revolution. They put an end to the general sabotage that we were confronted with after the October Revolution. . . . The Jewish elements were mobilized . . . and thus saved the revolution at a difficult time. It was only thanks to this pool of a rational and literate labor force that we succeeded in taking over the state apparatus."[78]

One reason, then, for the rise of Jews in science (as in at least some other literate professions) after the revolution was a rare confluence of circumstances. A new science infrastructure was hastily being built, so much so that the new soon greatly outweighed the old and established. In this situation, the power of an establishment scientific elite to stand in the way of young Jews and others wishing to enter science was negligible. This was all the more true because the sciences—most obviously physics but also biology and chemistry—themselves were rapidly changing in these years, undermining the august authority of old-school scientists. Jews, including the thousands of young Jews who trained in western Europe in the years prior to the revolution, were eager to enter Soviet sciences at a moment when the Soviet sciences were desperate precisely for such eager young scholars. New professions in search of professionals had met new professionals in search of professions.

None of this would have mattered, of course, if the postrevolutionary leaders of Soviet science—and, in a broader sense, Soviet society—had been resistant to allowing Jews to assume positions of leadership and influence. Few were; most were committed in word and deed to fashioning a genuine meritocracy. In 1930 William Horsley Gantt, an American physician who went to Russia in the twenties with the American Relief Administration and while there became an enthusiastic student of Pavlov, wrote an essay based on his travels, called "The Soviet's Treatment of Scientists."[79] In it he applauded Russian science for hiring and advancing scientists on the basis of merit and achievement rather than background, politics, or ideology.[80] The

institutionalized anti-Semitism of the past —quotas, forced conversions, and all the other blunt instruments that had kept Jews from positions of influence before the revolution—was outlawed with the revolution. Even the more pernicious and durable informal sorts of anti-Semitism lost much of their force. "Virtually all memoirists writing about Moscow and Leningrad intelligentsia life in the 1930s," Slezkine writes, "seem to agree that there was no anti-Jewish hostility and generally very few manifestations of ethnic ranking or labeling."[81] Vitaly Rubin, in time a world-famous sinologist, remembered his studies in a Moscow University, where half his classmates were Jews: "The Jewish question did not arise there. Not only did it not arise in the form of antisemitism; it did not arise at all. All the Jews knew themselves to be Jews but considered everything to do with Jewishness a thing of the past. I remember thinking of my father's stories about his childhood *heder* and traditional Jewish upbringing as something consigned to oblivion. None of that had anything to do with me. There was no active desire to renounce one's Jewishness. This problem simply did not exist."[82]

Scholars argue whether this state of affairs changed beginning in 1936, with the Great Purge. Most conclude, like historian David Priestland, that Jews as a group did not suffer special discrimination during the Purge or indeed at any time before World War II.[83] There is little disagreement with historian Zvi Gitelman's assessment of the first years after the revolution: "Never before in Russian history—and never subsequently—has a government made such an effort to uproot and stamp out antisemitism."[84] Part of the explanation for the great wave of Jews entering the sciences is the sudden, unforeseeable, and quite remarkable absence of official resistance to their doing so.

This absence of resistance did two things at once. For decades, there had been far more Jews who wished to attend university and enter learned professions than there had been places allotted to Jews. With the lifting of the barriers of quotas and official limitations on the professions which Jews could practice, these greater numbers streamed into and through the universities, with many ending up in laboratories and research centers. Once the dam was removed, the stream of Jews entering sciences swelled to what it would have been had the obstructions never existed. But the absence of resistance did something more. Removing obstructions not only allowed Soviet Jews to do what they in any case wanted to do, but it also *spurred* them toward those professions that were previously all but closed to them. The learned professions, the sciences among them, by virtue of being newly *available* to Jews, came to be seen as especially *meritorious* by Jews.

It is worth pondering the nature of this alchemy, through which the absence of a negative became a positive. To start with, it is important when speaking of Soviet Jews and their attraction to science to understand who these Jews were and who they were not. For one thing, these Jews were not always Jewish. Abram Ioffe himself converted in 1911 from Judaism (a religion he had not practiced since childhood) to Lutheranism (a religion he would never practice), in order to marry (as Russian law required). Few of his Jewish colleagues had reason to convert, but fewer still found meaning in any traditional formulation of Jewish religious practice, study, or belief. Yet their being Jews, or the children of Jews, was a fact of significance in their biographies. Jewish ethnicity remained a category that mattered, being one of the multitudes of accepted national minority groups in the Soviet Union, a "country of 189 peoples," as a 1934 travel poster had it.[85]

Following the revolution, the requirement to carry unified identification documents was rescinded, but when a system of internal passports was reinstituted in 1932, these new documents indicated peoples of "Jewish nationality." The nomenclature was important; from the start, Soviet social planners sought ways, not to wipe out Jewish identity, but to redefine it in "national" terms that fit more fully into the Soviet model of a state composed of many semi-autonomous nations or peoples.[86] It was for this reason that Yiddish schools, theaters, and presses were established, and that the far-flung Jewish autonomous oblast, or administrative region, was established in 1934, with Birobidzhan as its administrative center. The result of these efforts, over time, was paradoxical. The redefinition of Judaism as a nationality, with a language that only dwindling numbers of Jews wished to speak and a homeland that few Jews wished to occupy, drained Soviet Jews of their Jewish identity more than it sustained that identity. This was hardly viewed as a tragedy by most Jews; as Slezkine observed, "No other ethnic group was as good at being Soviet, and no other ethnic group was as keen on abandoning its language, rituals, and traditional areas of settlement."[87] But while the Soviet policy of identifying Jews as a nationality ultimately diminished Jewish identity, it at the same time contributed to the durability of, at the very least, the superficial identification of Jews as Jews. If it became harder in the Soviet Union to remain a Jew in anything but name alone, it at the same time became harder to escape remaining a Jew, at least in name only.

For Jews in these circumstances, science had special appeal. For one thing, it offered a pathway to full Soviet citizenship that was different from and superior to that offered by Jewish nationalism. Science in the first half of the twentieth century was almost universally associated with progress, improvement, and modernity. This was nowhere more true than in the Soviet Union, whose prophets and leaders—Marx, Engels, Plekhanov, Lenin, Trotsky, and Stalin—had each described their outlooks as "scientific" and had each praised science as a basic tool for reconstructing society on a sounder basis. When the American physicist Haroutune M. Dadourian returned from a long trip to the Soviet Union in 1930, he cannily observed that "science and the scientific method have assumed an importance in the minds of the Russian leaders second only to communism" itself.[88] Many Jews embraced science because science itself was seen as so valuable to the revolution while being so snuggly fit to the values of the revolution. To be a scientist was to demonstrate commitment to the Soviet ideal and to the Soviet Union. At the same time, to be a scientist in the Soviet Union, as elsewhere, demonstrated a commitment to universal ideals like the advance of human knowledge and the betterment of humankind. Science was an avenue whereby Jews could be superior Soviet citizens and superior human beings.

For some in the first decades after the revolution, this was part of the great appeal of science. And even when the corruption and oppression of the Soviet regime became too blatant to ignore (a moment reached by some Jews in the 1920s and by others only as late as the 1940s), science remained a profession through which one could remain a contributing Soviet citizen without being overly involved with the degrading ideological casuistry that was so often a part of Soviet life. Scientific institutions were not free of informants, loyalty tests, and Party apparatchiks, but these played less of a role in science than they did in any other intellectual profession.[89]

For Jews of the Soviet Union, science provided a profession that was hyper-modern, progressive, praiseworthy, and at once universalist and patriotic. It offered a high road into Soviet life while seeming to advance humanist values that a Jew one short generation from the Pale of Settlement might wish to see embraced within the Soviet Union and beyond. In this, the appeal of science for Soviet Jews was not terribly different from the appeal of science for American Jews, many of them too a scant generation from the Pale of Settlement. They all sought not simply to find their place in a society new to them, but also to remold these new societies in such a way that they might accommodate Jews.

When the great immigration of Soviet Jews to Israel began in the 1970s, waves of these Jews brought with them scientific training and skills, and expectations that these would somehow smooth their assimilation into their new home. These expectations were, after all, the same expectations that had drawn many of these Russian Jews, and their parents and grandparents, to science in the first place.

"Making a Land of Experiments"

Science and Technology in Zionist Imagination and Enterprise

IN THE SWELTER OF AUGUST 1960, 120 notables representing forty countries, mostly emerging nations in Africa and Asia, gathered at the Weizmann Institute in Rehovot, Israel, to attend the International Conference
on the Role of Science in the Advancement of New States.[1] Abba Eban had
conceived the event two years earlier in Washington, D.C., while he was Israel's ambassador to the United States. Since then, Eban had been appointed
president of the Weizmann Institute, then elected to the Knesset, and finally
chosen to serve as David Ben Gurion's minister of culture and education.
A man of uncommon stamina, Eban continued to hold all three posts. The
Rehovot conference was the rare event that captured at once a wide swathe
of Eban's interests and enthusiasms, engaging him as a politician, statesman,
educator, patron of Israeli science, and enthusiastic advocate of Zionism
and the still-young Jewish state.

After the conference, Eban submitted a report to the *Bulletin of the
Atomic Scientists* describing the meeting's aims and achievements: "The
theme of the conference was the capacity and duty of the modern scientific
movement to enrich the life of newly emerging communities. At the center
of its deliberations stood the two transformations which dominate the life
of our century—the rise of Asian and African peoples to independence and
the rapid progress of science and technology." The success of each of these
transformations, Eban said, depended on the other: "The history of our times
will be written largely by the two groups of men who came together for the
first time at the Rehovoth Conference—the statesmen of developing nations
and the leaders of scientific disciplines."[2]

Bringing these two groups together in a way that might produce freedom,

health, and democracy for developing nations, Eban continued, was a task for which Israel was uniquely fit.

> Our country stands at a crossroad—not only in geography, but also in the world of ideas. We are, by the fortune of history, a member of the modern world of science and technology. . . . We are also one of the new states of the international community, a partner in the modern enterprise of national liberation. We thus stand in simultaneous kinship to the scientists and to the representatives of new states assembled at Rehovoth. The fabric of Israel's history has a single unifying thread—a constant belief, not always easy to sustain, in the positive direction of human history, and in the responsiveness of men, when challenged by great issues and lofty ideas.
>
> Strong currents of passion still sweep across the awakening continents, threatening to submerge liberties hardly won and deeply cherished. It was a moving experience for us, at such a time, to set the stage of an international assembly consecrated to the pursuit of truth in the service of man's expanding welfare and enduring peace.[3]

Israel's singular position as both a scientific powerhouse and an emerging nation made it a natural bridge, Eban believed, between the developed world and the developing world. It was what gave Israeli scientists the authority to teach the leaders of newly established African and Asian nations from experience. This was obvious during the conference when, Eban boasted,

> Dr. Zvi Tabor of Jerusalem showed a "sound pond" and solar-energy boilers which indicated early possibilities of turning Asian and African sunshine to practical account. . . . Mr. Zarchin of Israel gave varying prognoses of the desalinization of sea-water. . . . Mr. Aaron Wiener of Israel's Water Planning Organization discussed the agricultural problems of widely divergent climates. . . . Professor Saul Adler of Jerusalem proposed broad regional planning for the elimination of the tsetse fly. . . . [and] population regulation [was] outlined by Professor Shelesnyak of the Weizmann Institute.[4]

Conference participants were also taken around the country on tour buses to visit examples of highly productive industrialized agriculture on collective settlements, desalination plants, big infrastructure projects, and other examples of small and large successes produced by Israeli scientific planning and development. The "concluding discussions" of the conference were devoted to "assistance offered by Israel." Yohanan Ratner of the Technion pledged "to provide training at our school on the secondary level for technicians and foremen [from developing countries], and in addition to train a smaller number of engineers for four years, on the average, giving them the opportunity to take part in the scientific work conducted in our laboratories to the extent that they are able." Gehard Schmidt of the Weizmann Institute declared the school's "readiness to accept graduate students from the new states. . . . We have immediately available a number of scholarships for the maintenance of students from the African and Asian continents." Saul Adler of the Hebrew University of Jerusalem arose to say that "students from Africa and Asia will also find [in Jerusalem] laboratories equipped for almost every branch of micro-biology, a subject which may at first appear to be academic as far as some of the new states are concerned but which really has the widest applications for animal and human welfare and also for industrial development." Amos Maor of the Histadrut labor union offered to provide six-month courses in the union's Afro-Asian Institute of Labor Studies "for 60 participants from African and Asian countries." A representative of the Israel Atomic Energy Commission offered to open the "Israel radioscope training center" to students from new countries. Another participant from ORT, the Organization for Rehabilitation and Training, offered technical training. The Institute for Fibers and Forest Products offered "specialized training of technicians, engineers and chemists in these sciences and technologies." Eban, who chaired the session, saw things this way: the offers of assistance from Israel, "however small its size and modest its resources," at least set an example for others. "I hope that these statements will have an exemplary effect upon other governments, and especially other scientific institutions."[5]

It was with satisfaction that Eban reported the remark of the Reverend Solomon Caulker, vice-principal of a college in Sierra Leone: "I came in darkness, but I leave in light."[6] "The image of sudden illuminations aptly sums up the lasting impression of this encounter between the science and statescraft of this century," Eban wrote.[7] Caulker's evocative image of light emanating from the campus of the Weizmann Institute to the dark recesses of Africa captures the complexity of feelings that many Israelis, like Eban, had toward science and its place in Israeli society—how attitudes toward science affected Israelis' self-image, how Israel was viewed from beyond its borders, and countless day-to-day efforts to build and maintain the country. By 1960, science and technology were an important part of Zionist ideology, psychology, politics, and praxis. Indeed, they had been since the very start of the Zionist project, almost a century before Abba Eban convened the Rehovot conference.

Science and Early Zionist Thought

The notion that Israel's fate is tied up with science is far older than the state itself. In 1892, Russian journalist and activist Elhanan Leib Lewinsky published a popular utopian novel called *Masa' le-Erets Yisra'el bi-shenat tat (2040)* (Voyage to the Land of Israel in the year 5800 [2040]). Lewinsky had dropped out of medical school to devote himself to the cause of Jewish nationalism after the wave of pogroms in 1881,[8] even traveling for several months to Palestine. The country he described, in a safely distant future, was far more Western and developed than the country he observed firsthand in 1881:

> In all the history of new settlements, it was unseen and unheard of for
> a country to be based on a foundation of justice. One hundred and fifty
> years ago, our forefathers were naïve enough to genuinely believe that it
> is possible to so found colonies and build a country. . . . Everyone now
> understands that Darwinian theory, with its iron-clad rule of the war for
> survival, is especially appropriate for new colonization. Here we see how
> the strong prevails and succeeds and inherits the land, and the weak falls
> plundered in the war, and will leave behind no trace, even if he has all the
> support in the world. . . . In the Land of Israel [were established] for the
> general good: academies, schools, museums, libraries, parks, tramways,
> steam and electrical ships, medical clinics, baths, canals and the like for
> the public good. . . . All of the Land of Israel is now perfectly established.
> Cobblestone and iron tracks connect village to village, there are parks and
> orchards in every settlement, houses of worship and study and learn-
> ing are well fashioned, there are extravagant libraries, hospitals, clinics,
> higher and lower springs, quarries, steam and electric ships, until the land
> before us became a Garden of Eden.[9]

The Land of Israel of the future that Lewinsky goes on to describe is a workers' paradise and a model of distributive justice, but it is equally the most advanced of European technocracies. Jaffa, where his protagonist makes landfall, is blessed with modern lighting and efficient, free tramways, and hosts an academy of nautical studies. (Inland, he visits an academy of geological studies, and Jerusalem has a university of high renown.) Visiting a professor he finds "a great treasure trove of books, most from the recent century, with much literature on agriculture and botany in Hebrew, Arabic, French, English, and German."[10] After describing the copious private property that the citizens of the Land of Israel have at their disposal, Lewinsky remarks: "But what is private property compared to the national government's property: all the rail tracks, steam ships, telegraph, telephone, phonographs, airships, coal and metal mines: all belonging to the government, and everyone is free to enjoy them?"[11] So advanced and successful is the country Lewinsky describes that it eclipses the capitals of Europe. *Voyage to the Land of Israel* ends with this: "In the past, our forefathers traveled to Jerusalem by way of Paris, and because they found themselves in Paris, forsook continuing to Jerusalem. Now the order is turned. We travel to Paris by way of Jerusalem, and when we come to Jerusalem, we forsake Paris. How times have changed!"[12]

The most famous and influential of early Zionist ideologues and activists, Theodor Herzl, also foresaw a Jewish homeland that would surpass the technological and scientific wonders of Europe. His was what one scholar called a "scientific messianism."[13] Herzl wrote in his 1896 manifesto *Der Judenstaat* (*The Jewish State*) that "the founding of a Jewish State, as I conceive it, presupposes the application of scientific methods. We cannot journey out of Egypt today in the primitive fashion of ancient times."[14] As he neared the conclusion, Herzl grew elegiac:

> The word "impossible" has ceased to exist in the vocabulary of technical science. Were a man who lived in the last century to return to the earth, he would find the life of today full of incomprehensible magic. Wherever the moderns appear with our inventions, we transform the desert into a garden. To build a city takes in our time as many years as it formerly required centuries; America offers endless examples of this. Distance has ceased to be an obstacle. The spirit of our age has gathered fabulous treasures into its storehouse. Every day this wealth increases. A hundred thousand heads are occupied with speculations and research at every point of the globe, and what any one discovers belongs the next moment to the whole world. We ourselves will use and carry on every new attempt in our Jewish land; and just as we shall introduce the seven-hour day as an experiment for the good of humanity, so we shall proceed in everything else in the same humane spirit, making of the new land a land of experiments and a model State.[15]

In his utopian novel *Altneuland* (*Old-New Land*), Herzl gave substance to these abstractions, describing a society more mechanized and technologically advanced than any the Jews left behind in Europe. European visitors, returning to a Palestine that not long before was hopelessly primitive, are amazed at the improvements engineered by Jewish immigrants to the region:

> They had to halt at a railway crossing because a train was due. It appeared presently, rushing southward at great speed. When the visitors

remarked that the locomotive had no smokestack, they were told that this line, like most of the Palestinian railways, was operated by electric power. There was one of the great advantages of having begun from the beginning. Just because everything here had been in a primitive, neglected state, it had been possible to install the most up-to-date technical appliances at once. So it had been with the city planning, as they already knew; and so it had been with the construction of railways, the digging of canals, the establishment of agriculture and industry in the land. The Jewish settlers who streamed into the country had brought with them the experience of the whole civilized world. The trained men graduated from universities, technical, agricultural and commercial colleges had brought with them every type of skill required for building up the country. The penniless young intelligentsia, for whom there were no opportunities in the anti-Semitic countries and who there sank to the level of a hopeless, revolutionary-minded proletariat, these desperate, educated young men had become a great blessing for Palestine, for they had brought the latest methods of applied science into the country.[16]

The image of a state that is a marvel of technological and scientific planning and development was a part of almost every early attempt to imagine Jewish sovereignty. In 1899, an orthodox American Jew named Henry Periera Mendes published his own utopian novel, *Looking Ahead*, which finds in a future Jewish state in Palestine the salvation of all humankind. Mendes was an intriguing figure who resists easy characterization. He was an orthodox Jew, and at the same time he was a founder of New York's Jewish Theological Seminary. In the society he envisioned, the rule of the rabbis was firm. Yet even in his utopian theocracy, Jewish sovereignty was recognizable by its technological development and scientific sophistication: "Roads were made, villages were rebuilt, enlarged into towns, watercourses were constructed, fields were planted, and the growth of the towns into cities bade fair to rival the miracles of Chicago and San Francisco. . . . Factories sprang into existence. Immense coal fields were found toward Euphrates, petroleum to the south, metals in the Midian Hills. Railways, as if by magic, branched from Jerusalem, Damascus and Lebanon, and met railway systems of other lands. . . . Palestine [was] now recognized as the future emporium for the world."[17] Mendes' book ends with this coda, an encomium to Jewish smarts and the light they might bring unto the nations: "No need to describe how the University of Zion became an influence for good, so that human happiness became measurably nearer, and humanity learned that restoration of Palestine to the Jews meant really the restoration of all men to the love of the common Father of all. No need to speak of the world's progress since then. No need to speak of what we all taste to-day—universal peace, universal brotherhood, universal happiness."[18]

Similar themes continued to appear in the fiction of Jewish futurology, resurfacing again in 1922, well after the lurching Zionist settlement of Palestine had produced a reality that rendered utopian novels more fantastic and less plausible than they had early been. Haim Shalom Ben-Avram, in *Kommemuyot* (Independence), described a Jewish state easily accessible by air or sea, fashioned into a paradise by Jewish energy and ingenuity. Radio stations were built. Great factories with thumping machines were established. Train tracks stitched together the cities, towns, and villages. When, inevitably, the country faced a shortage of power, "it occurred to a young engineer" to dam the Jordan.[19] Wind turbines were added. Jews brought with them knowledge from the many places around the globe they had lived, and they produced for themselves the most advanced of nations.

In all of these attempts to foresee the future Jewish state, practical scientific and engineering advances were inseparable from those in social engineering. In each case, new knowledge and new machines were part of engineering a new society and a new Jew. Much was made of the new forms of settlement, new economic arrangements, and new social norms that would evolve in line with rational principles. As reflected in the earlier visionary novels, the homeland was seen not just as a product of the most advanced technologies, informed by the most advanced science. It was a full-blown technocracy, in which scientific principles, interpreted by men schooled in scientific management, scientific planning, scientific agriculture, and so forth, would have ample influence over how the country was developed and administered.[20]

From Vision to Practice

This vision of a shining technocracy in the Levant was not the sole province of novelist-dreamers. It was an everyday part of the thought and rhetoric of Zionists grounded in the here and now. From its earliest years, "Zionism" described a group of diverse ideologies, outlooks, and programs under great centrifugal pressure. There were political Zionists, practical Zionists and synthetic ones, secular and religious and spiritual Zionists, cultural Zionists, labor Zionists, socialist Zionists, collectivist Zionists, revisionist Zionists, general Zionists, and more. The succeeding waves of immigration in the first decades of Zionism—a First Aliyah, and a Second, Third, Fourth and so on—each had a unique collage of characters and concerns. The faultlines between Zionists of different backgrounds, beliefs, and ideologies were multitudinous, and across them there were regular eruptions of anger and anxiety. Across this latticework of divides there were relatively few beliefs about which Zionists shared consensus. The notion that Jewish settlement in Palestine should proceed in a scientific fashion, and that the land would be remade by Jews alive to the power of new scientific methods and new technologies, was one of these rare beliefs. It spanned chasms of ideology, background, language, and lifestyle.[21]

Pick almost any influential Zionist, and you will find that he or she held one version or another of this belief. Vladimir Jabotinsky—the revisionist ideologue and tireless Zionist activist, politician, and diplomat—insisted that critics were wrong to call Herzl's novels utopian and romantic; while they were literary works of the highest order, aptly assuming some poetic license, they were at heart realistic assessments of what Zionism could produce. Jabotinsky wrote glowingly of *Altneuland* in a 1905 essay called "Doctor Herzl": "The picture Herzl painted is the picture of a social regime unparalleled in its progressiveness." It was a vision that Jabotinsky (who admitted affection for the futuristic yarns of Jules Verne) himself shared.[22]

Jabotinsky's enthusiasm for Herzl's vision of a liberal technocracy would perhaps be surprising had it not been shared by so many early Zionists of so many varied backgrounds and worldviews. One of these was Max Nordau, the physician-intellectual who persuaded Herzl to initiate the Zionist Congresses that became a hub of the Zionist movement in its first decades and a venue for efforts to translate ideas into policies and policies into practical projects and programs. Nordau included in his *Interpretation of History* a lyrical tribute to how science and technology had moved humanity to a new standing. Consider, he wrote, the "gap between the little oil-lamp and pinewood torch and the electric light! Between the kindling of fire by the tinder and by a match! Between traveling on foot, horseback, or on a raft, and in the electric train or turbine steamer! Between sending a message on foot or by means of telegraph and telephone! Between the club and axe of stone and the revolver, machine gun, torpedo, and armoured cruiser! Why prolong a recital that every educated man can complete for himself? Here, progress is undeniable."[23]

The effects of this progress, Nordau continued, go much further than the comforts they provide: "Progress in knowledge permits all the resources of nature that can be used by man to be more profitably employed, the evils and dangers that threatened him to be more frequently avoided, pleasure to be increased, discomfort lessened, and the average duration of life to be prolonged. The immediate effect of increased knowledge is purely utilitarian and biological. Indirectly it is psychological and moral. It increases self-reliance in man, and gives him a rising sense of his own dignity. It rouses resistance to selfish domination, tutelage, exploitation."[24] Nordau believed that it was precisely this rising sense of dignity and resistance to domination that accounted for the nationalist awakening of Jews (though not Jews alone). This progress, the moral progress that rises alongside scientific and technological progress, is at the heart of Zionism and at the heart of the modern condition itself.

Historian Derek Penslar put it bluntly: "The Zionist movement revered technical expertise as an essential tool for the construction of a Jewish homeland," and while there were lonely exceptions to this rule, they were surprisingly few.[25] Speaking before an "emergency Palestine economic conference" in Washington, D.C., in November 1929, U.S. Supreme Court justice and Zionist leader Louis Brandeis observed that "the Jewish pioneers demonstrated that it was still possible to make Palestine into a land flowing with milk and honey and with much besides. Touched by intelligent effort supplemented by science, it began to bloom almost as a miracle."[26] Earlier, Brandeis had said that American Jewish doctors and scientists had made possible the notion of a Jewish state when they "undertook to make health possible in Palestine. And it really was not a difficult problem. For the lack of health was largely due to malaria. Happily, science enables us to grapple with this disease which had devastated many countries of the world for thousands of years. We know how to rid a land of it."[27] As a matter of principle and practice, Brandeis maintained, Jews would capture the Land of Israel through science and technology.

This view had become, by Brandeis' time, a commonplace. In 1928, when Irma Levy Lindheim replaced Henrietta Szold as the president of Hadassah, her panegyric to the organization's founder was that "Henrietta Szold applied the scientific method in organizing. . . . She reduced the general Zionist idea to a particular part of its program and then proceeded to develop bit by bit the instrument with which to construct this part."[28] As a result, she wrote soon thereafter, "New life and hope are being brought to the East in the standards Hadassah is setting. It is amazing to see the most modern and hygienic methods of handling milk in practice here, in the midst of unspeakable ignorance and filth. The fairy wand of science is lighting up the dark corners of the earth, and the hand of woman is holding up the torch. Hadassah is certainly the Mother of Palestine today. She is tending to her children and healing their ills. She has reduced the infant mortality and blindness from trachoma to a very considerable degree."[29]

Arthur Ruppin, the gifted Zionist functionary who headed the World Zionist Organization's Palestine Bureau (which acquired land for Jewish settlement) and who in time joined the faculty of the Hebrew University as a sociologist, was no less impressed with the advance of Western medicine that came to the Levant with the Jewish immigrants: "The facts that a large hospital attached to the Hebrew University has been built, that a Cancer Research Institute has been established, and that many famous physicians, especially from Berlin and Vienna, have settled in Palestine in recent years, are likely to make Palestine the medical centre for the Jews of the whole world, and also for non-Jewish patients in the Near East."[30]

Advanced medicine was only a small part of the scientific modernization wrought by the influx of Jews.[31] There were also advances in "sanitation, afforestation and land registration [which] were organized on a modern basis." Thanks to "a huge influx of Jewish capitalists and experts," Ruppin wrote, "considerable cultural and economic advance was achieved in the country. . . . A modern harbour was constructed at Haifa. Jewish initiative led to the establishment of a hydro-electric plant providing the country with power and light. From 1932, while nearly all other countries of the world were suffering from a severe economic crisis, Palestine experienced an unprecedented boom, as a result of the influx of Jewish immigrants and Jewish capital."[32]

For Ruppin, as for so many others, the notion of the return of Jews to the Holy Land was inseparable from an account of the advanced science they brought with them and the progress this science would bring. This notion that contemporary science and newly devised technologies could aid Jews in settling Palestine and in laying persuasive claim to the land was put into practice almost at the start of Zionist immigration. As early as 1899, the German botanist and expert in tropical plants Otto Warburg undertook a research junket to Palestine (with stops in Cyprus and Turkey as well) to investigate the suitability of the climate for growing new agricultural products.[33] Warburg had met Herzl a year earlier and in time supplied him with botanical background information about Palestine for *Altneuland*, which Herzl was then writing. Employing his science in the service of a colonial project was not new to Warburg; while teaching at Berlin University he had helped found the Institut für Kolonialwirtschaft (Institute for Colonial Science) and wrote monographs about plantation production of cocoa, coffee, and rubber, core crops of German colonial agriculture.[34] Beginning in 1897, Warburg served as publisher and editor of the journal of the German Colonial Agriculture Committee, *Der Tropenpflanzer*.

After investigating Palestine on his first visit, Warburg was persuaded that botany and agronomy could help Jews develop Palestine in much the same way that these sciences had helped Germany develop her African colonies. In 1903, Warburg received an opportunity to test this persuasion when he was appointed to establish and direct the World Zionist Organization's Commission for the Exploration of Palestine. Outstanding among his colleagues on the commission were Selig Soskin, a Russian Jew who in Berlin had earned his doctorate in agronomy; Joseph Triedel, a German hydraulic engineer who had earned his degree in Bonn; and Aaron Aaronsohn, a Romanian agronomist who had spent his childhood from age six in the Jewish agricultural colony of Zichron Yaakov and had studied in Montpellier. By April 1904, the four men had come up with a proposal that Warburg and Soskin presented to the WZO Executive Committee, to establish a training farm, an experimental station, and an experimental cooperative settlement in Palestine.[35]

Although Warburg's proposals found a mixed reception—there were those who found them out of step with revolutionary collectivism—the schemes reflected an approach that continued to find traction as Jewish settlement of Palestine continued. By 1911, Aaron Aaronsohn had established, with the support of American Jewish philanthropists, a Jewish agricultural experiment station in Atlit.[36] Aaronsohn had achieved international fame four years earlier with, as *Science* reported it, "his discovery of the long-sought wild prototype of wheat." He raised cash from American Jewish philanthropists and politicians—many of whom (like Julius Rosenwald, Samuel Fels, and Jacob Schiff) also supported and promoted American science and medicine—explaining that his labs would draw experts from around the world and "go far towards introducing American methods in the study of agricultural problems throughout the whole Mediterranean region."[37] Soon other agricultural research centers were established as well, with the aim of enabling Jewish pioneers to bring Western science to the Levant and, conversely, to bring insights gained from the unique research opportunities offered by the Levant to Western science.[38]

Others, of a more bookish bent, sought to establish universities with the same aim. The idea of a Jewish university in Palestine was first proposed in 1882 by a German mathematics professor (and ordained rabbi) named Zvi Herman Shapira. Shapira wrote in the Hebrew paper *Ha-Melits*:

> We must take care from the very beginning of the establishment of settlements in the land of our forefathers to establish in the center of these settlements a great house of learning, from which wisdom will emanate, wisdom and morality for all the house of Israel. This house shall be divided in my opinion into departments: (1) theology, (2) theoretical sciences, and (3) practical sciences. . . . The theoretical department will teach

natural sciences, geometry, mechanics, astronomy, etc. (about which it was written, "This is your wisdom and understanding in the sight of the peoples" [Deut 4:6]). And the practical department will teach chemistry, botany, geology, architecture, and agriculture.[39]

The first Zionist Congress, meeting in Basel in 1897, discussed Shapira's idea approvingly; and at the fifth Congress, meeting in Basel in 1901, first steps were taken to make it a reality. Following the meetings, chemistry professor Chaim Weizmann, philosopher Martin Buber, and editor and publisher Berthold Feiwel collaborated on an influential pamphlet called "A Jewish Institute of Higher Education," advocating the establishment of

> a place for Jewish youth prevented from learning a profession in the lands of their birth, and for whom the gates of science are closing in their faces. . . . Important Jewish scholars, who are deprived because of their backgrounds . . . would find in it a place in which they could devote themselves entirely to science, and also entirely to their people. . . . This undertaking, were it to exist, would serve our nation as a proud proof of its living and creating power, and this proof would provide the strength and confidence in still greater national achievement.[40]

At about the same time, Herzl proposed to Abdul-Hamid II, sultan of the Ottoman Empire, the establishment of a Jewish university in Jerusalem that would serve subjects of the empire of all backgrounds and religions. The sultan dismissed the idea as impractical. Still, a fund-raising apparatus was devised by the World Zionist Organization and set into motion; and over the first two decades of the twentieth century, and especially beginning in 1914, land and buildings on Mt. Scopus were purchased as the site of a future university. In 1918, after the First World War had dismantled the Ottoman Empire and established the British as the administrators of Palestine, Weizmann—who was by this time internationally famous as a scientist and, in different circles, as president of the British Zionist Federation—laid the cornerstone of the Hebrew University in Jerusalem, declaring that the university would strive to bring blessings not just to Jews but to all nations.

There was touching sincerity in Weizmann's faith in science as a bridge between cultures. It was his own genius for chemistry that had lifted Weizmann, son of a struggling timber merchant who had studied in *heder*, to world fame and to the ballrooms and boardrooms of Europe. Weizmann had also demonstrated that science can be of immense and immediate practical value. The discovery for which he achieved fame was a process for producing acetone, a substance crucial to the mass-production of the explosives that the United Kingdom employed in unprecedented amounts during the Great War.[41] Weizmann understood that science in the modern age was both an indispensable tool and a portal into Western society, and both aspects were of great value to Jews trying to establish a national home in Palestine. It was Hannah Arendt who best described Weizmann's attitude: "For him science is not the eternal search for truth but the urge 'to make something practical,' an instrument for a well-defined task: the building of Palestine most of all, but also the possibility of that financial independence to which he owes so much of his political success, and, last not least, his unsurpassable entrance ticket to the international world."[42]

In 1921, Weizmann and Einstein barnstormed the United States, raising money for the new university.[43] "They don't need me for my abilities but because of my name, whose luster they hope will attract quite a bit of success with the rich kinsmen of Dollar-land," Einstein wrote to his friend, the Nobel laureate chemist (and formerly Jewish) Fritz Haber.[44] There was bite to Einstein's ridicule, but it reflected a truth worth pondering. It was not a coincidence that two great Jewish scientists (joined by the engineer Menachem Mendel Ussishkin and Ben-Zion Mossinsohn, the principal of Gymnasia Herzlia) were canvassing for cash to fund a Jewish university in the new Jewish settlements of Palestine. For obvious reasons, there was sense in scientists advancing the cause of a university. But as one may gather from the thunderous reception they enjoyed as they barnstormed the country ("Vast throngs of New York Jews turned out to greet Professor Chaim Weizmann, discoverer of T.N.T. and president of the World Zionist Organization, and Professor Albert Einstein, famous savant" was the headline of the *Jewish Independent* on April 8, 1921), something else was afoot as well.[45] Weizmann's and Einstein's reputations had been earned for their surpassing *human* achievements, and they were now conferring some of the "luster," as Einstein put it, of these achievements upon the Zionist project, and perhaps upon Judaism in general. Rabbi Abba Hillel Silver had traveled from Cleveland to New York to be on hand for the arrival of the scientists, and the honor of introducing them at a gala event at the Metropolitan Opera House fell to him. "We greet, all of us," he began, "that man, that intellectual Titan, who has again given evidence through his labors and his achievements of the intellectual leadership of the sons and daughters of Israel throughout the world, Albert Einstein."[46]

It was not only Jews who saw that the luster of scientific success added gleam to Zionist aspirations. President Warren Harding was unable to meet the delegation in New York, and sent a letter of regret praising the scientists: "Representing as they do leadership in two different realms, their visit must remind people of the great services that the Jewish race have rendered to humanity."[47] There were also those who received Weizmann and Einstein with ambivalence, finding something unseemly in the blurring of the border between science and Zionism. Chemist and science journalist Edwin Emery Slosson wrote a long essay called "Einstein's Reception" in the Congregationalist journal *The Independent*. The piece opened with an appreciative tone: "True science is above all barriers of nation, race or sect. Einstein himself, although he is an ardent nationalist and comes to America as an advocate of Zionism, is nevertheless a sincere internationalist and welcomes all efforts to reestablish the world commonwealth of science." But Einstein's own universalist predilections, Slosson continued, hardly justified his yoking his scientific achievements to Zionist aims:

> It is unfortunate that Einstein should make his first appearance in America as a Zionist instead of a scientist. He would have done more for Judaism in general and for Zionism in particular if he had come to America, like Bergson, as one of the great thinkers of the modern world whom all America delighted to honor rather than as a leader in a separatist movement of a race. But so long as there are some who hold the Jews as a whole responsible for those of their race who have itching palms or dirty fingernails, we must expect them to show in return a disposition to monopolize their men of genius.[48]

There was substance to Slosson's surmise that the appeal of a "Hebrew" University in Palestine was, to some, that it would allow Zionists to claim (if not, in the end, "monopolize") their men of genius. The motivations for establishing a university were many and complicated; a Hebrew University served as a symbol of many and varied things to many and varied people, and it also promised practical benefits of different sorts to different people.[49] These diverse motivations were on display at the official opening of the university, which took place in an amphitheater on Mount Scopus on April 1, 1925. It was a festive affair, attended, as the volume published in commemoration of the affair reports, by "some six or seven thousand persons including many visitors who traveled thousands of miles to be present at the ceremony."[50] There were many speeches, some blandly congratulatory, some expansively poetical. Sir Herbert Samuel, the British high commissioner in Palestine, predicted that "in this House of Wisdom, there will be studied and taught the most ancient literature and the most modern science, side by side."[51]

Weizmann was the one most people came to hear:

What we are inaugurating to-day is a Hebrew University. Hebrew will
be the language of its schools and Colleges. But a University is nothing
if it is not universal. It must stand not only for the pursuit of every form
of knowledge which the mind of man embraces, but also for a commonwealth of learning freely open to all men and women of every creed and
race. Within the precincts of these Schools political strife and division
cease and all creeds and races will, I hope, be united in the great common
task of searching for truth, in restoring to Palestine the thriving civilization which it once enjoyed, and in giving it a place of its own in the world
of thought and learning. Our University would not be true to itself or
to Jewish traditions, if it were not a house of study for all peoples and
more especially for all the peoples of Palestine. Conceived in this spirit,
and animated by these ideals, the University has before it, if our hopes
are realized, a future pregnant with possibilities, not only for the Jews of
Palestine, but also for the awakening East and for mankind at large.[52]

Weizmann declared that the new university would

win its spurs and build up its reputation by the distinctive value of its
contributions to the common stock of knowledge. We have begun with a
group of institutes for advanced research in those branches of science and
learning for which Palestine offers particularly congenial soil. . . . Three
such institutes [will be] devoted respectively to chemistry, to microbiol-
ogy, and to Jewish and Oriental studies, and before these celebrations are
concluded, we shall lay the foundation stone of an Institute of Physics and
Mathematics to be associated with the name of Einstein.[53]

In Weizmann's views alone, one finds a collage of justifications for the
university. It will serve humanity. It will serve Palestine. It will promote the
flourishing of Jewish culture. It will allow Jews to spark the flourishing
of Western culture. It will shine a klieg light on the genius of Jews, and it
will allow Jews to forge lines with "the awakening East" and "mankind at
large." This complex of views was widely shared by Jews of varied outlooks.
"Whether you are a Zionist, a non-Zionist or even an anti-Zionist," Louis
Gershenfeld wrote in the conclusion to his 1934 *The Jew in Science*,

you must grant that the rise of the Hebrew University in Zion will eventually more than compensate Jewry for any blunders (if you honestly believe
they are blunders) in its attempt in the colonization of Palestine. . . . It
is the Hebrew University in Palestine which must serve as the modern

agency, yes the powerhouse and the experimental station and the laboratory for revitalizing the Jewish heritage, for aiding in the solution of many Jewish problems, and to serve as a permanent fixture constantly supplying life and force not only to Jewish activities the world over but to scholarship, culture, science and to humanity at large.[54]

The Hebrew University was not the only center for scientific and scholarly research, teaching, and learning to arise in those years. In 1912, the cornerstone had been laid in Haifa for the Haifa Technical Institute, or Technicum, which was first conceived as a high-level technical secondary school. By the time the institute opened in 1924, after fundraising difficulties and the war slowed its construction, it was an institute of higher education. In its first year, twenty students enrolled to study civil engineering and architecture. Divisions of mechanical and electrical engineering were soon added, and the student body grew rapidly.[55] In 1934, the Daniel Sieff Research Institute was founded in Rehovot, quickly establishing itself as a center for chemistry research. It was here that Chaim Weizmann established his laboratory. Beginning in 1944, to mark Weizmann's seventieth birthday, the Sieff Institute was gradually expanded and renamed the Weizmann Institute of Science (which was finally formally incorporated in 1949).[56] By the mid-1950s, when new universities were hastily being planned for Ramat Gan, Tel Aviv, Haifa, and Beer Sheva, about 6,300 students were studying in institutions of higher learning in Israel; of these, about 4,000 students, or 63%, were studying natural sciences, medicine, agricultural sciences, and engineering.[57]

Universities, research centers, and experimental stations were the sites of formal science in the Jewish settlement in Palestine, and their importance only grew with time. But most residents of the Yishuv, as Jewish Palestine was known before the establishment of the state of Israel, came to experience science from well outside the laboratory and the university classroom. Farmers on socialist *kibbutzim* and cooperative *moshavim* were visited regularly by agricultural advisors who brought with them new agricultural techniques and, often, newly developed seeds for crops and feeds for livestock. Increases in farm production were publicized and celebrated in Palestine with enthusiasm, symbols of the power of Jewish sweat and science. "Where twenty years ago the soil was either marshland or just scratched with primitive plow," the narrator of the 1935 movie *The Land of Promise* —the first "talkie" produced in Palestine—intoned with baritone authority, "the modernized machinery and modern methods of the Jews obtain a maximum yield."[58]

In farms, towns, and cities, scientific medicine and public health assumed a more intimate place in people's lives. After the First World War, all new immigrants to Palestine were required to undergo medical examinations in their ports of exit, and then again when they arrived. Medical authorities in the *yishuv* grew more aggressive in monitoring public health, administering vaccinations, mandating prenatal care, and offering nutritional and other health advice, especially to new immigrants from underdeveloped areas. Schools taught scientific diet and hygiene in an effort to bring to Western standards children raised by "primitive" parents. "We are spreading culture," was how one nurse explained their mission to a group of young "health scouts" in 1931.[59]

Science and technology entered everyday life through architecture as well. One of the most celebrated settlements in Palestine was Nahalal, designed in 1921 by architect Richard Kauffmann as a pattern of perfect concentric circles radiating outward: in the center were "public" buildings like the school, auditorium, and warehouses, which were surrounded in turn by homesteads, then greenhouses and cowsheds, and then fields.[60] This form of scientific modernism had echoes that resounded at almost every level of planning.[61] In 1919, the Scottish biologist-*cum*-urban planner Patrick Geddes was hired with his son-in-law, architect Sir Frank Mear, to plan the Hebrew University, having impressed a group of prominent British Zionists that included Weizmann.[62] Geddes' appeal owed, in part, to his confidence that his planning reflected the insights gained as a Spencerian biologist, and especially to his theory of "reciprocal accommodation," for which he had gained fame and a letter of enthusiastic admiration from Charles Darwin himself.[63]

When Geddes returned to Israel in 1925 for the opening of Hebrew University, he was pressed into service producing a gridwork master plan for Tel Aviv. As Geddes worked to ensure that the city developed in the rational and orderly fashion he advocated, European-trained architects in Tel Aviv began to shape the city, building by building, in Bauhaus, form-follows-function "International Style." Staatliches Bauhaus was a German school of fine arts, craft, design, and architecture that flourished for 14 years between when it opened in 1919 and when the Nazis squeezed it closed in 1933. During this brief period it moved from Weimar to Dessau to Berlin, under the successive leadership of three remarkable architects, Walter Gropius (from 1919 to 1928), Hannes Meyer (1928–30), and Ludwig Mies van der Rohe (1930–33).

Each of the men, and their colleagues and students, saw themselves as doing nothing less than placing art and architecture on a new foundation appropriate to the age of science and technology.[64] "The Bauhaus workshops are essentially laboratories in which prototypes suitable for mass production and typical of their time are developed with care and constantly improved," Gropius wrote in 1926. "Only by constant contact with advanced technology, with the diversity of new materials and with new methods of construction, is the creative individual able . . . to develop from that a new attitude to design, namely: Determined acceptance of the living environment of machines and vehicles."[65] Meyer wrote that "building is not an aesthetic process." Rather, any new construction is "a product of industry and as such is the work of specialists: economists, statisticians, hygienicists, climatologists" as well as experts in "norms" and heating techniques. The architect is an artist, but he "is becoming a specialist in organization. . . . Building is only organization: social, technical, economic, mental organization."[66]

It was to this school, and this outlook, that the architects who planned Tel Aviv were drawn in growing numbers.[67] Four architects who came to have towering influence and towering reputations—Shlomo Bernstein, Shmuel Miestechkin, Arieh Sharon, and Munio Weinraub—traveled to Germany to receive training at Bauhaus and then settled in Tel Aviv.[68] They, and a larger number of colleagues who became acquainted with the International Style in their studies in Paris, Brussels, Zurich, or elsewhere, adopted the scientific, modernist approach with an enthusiasm unmatched anywhere else. By the start of the Second World War, Tel Aviv had become the world's leading exemplar of the pinnacle of modernist, scientifically inflected architecture.[69]

Science informed not only the efforts to inhabit the city but also the efforts to conquer the wilderness. Geographers and cartographers set out to map and document Palestine, replacing Arabic names with Hebrew names as they did, and, what's most extraordinary, some of these men became national heroes.[70] The introduction to a children's book, *The Pioneers: The Nature Researchers of the Land of Israel*, captures this well in its epigraph: "A hundred years ago, our land was unknown. The flowers and the trees that bloomed in its fields and hills, the birds that swooped from branch to branch, the wild animals that wandered on its paths were a sort of mystery. In 1863, a British research expedition arrived in the land of Israel. . . . [Then] came Jewish researchers—Aaron Aaronsohn who discovered the prototype of wheat, Ephraim ha-Reuveni the researcher of the country's plants, and Israel Aharoni the zoologist and many others as well."[71]

These "Jewish researchers" became celebrities in their own right. Israel Aharoni, for example, was engaged in 1908 by Sultan Abdul-Hamid II to perform zoological surveys of Palestine. Aharoni devoted the next thirty-eight years to cataloging the fauna of the Holy Land, collecting species, anointing them like Adam with Hebrew names, publishing scientific notices, writing Hebrew field guides to Palestine, and tramping into the wild with generations of schoolchildren. When he published his *Memoirs of a Hebrew Zoologist* near the end of his celebrated career, it quickly went through printing after printing.[72]

The appeal of figures like Aharoni was complicated and contradictory. The Jewish botanists, zoologists, geographers, and geologists who fanned out over Palestine in the first decades of Jewish settlement were at least two things at once. They were outdoorsmen—men of action, sunswept and ruddy of complexion, with dirty hands and torn work clothes. In this, they embodied the pioneering ideal, with its echoes of German Romantic rejection of sallow intellectualism. But at the same time, they were men of science, bringing to Palestine for the first time the system, the promise, and the progress so easily associated at this time with Western science. And through science, they seemed to offer a justified, if not necessarily just, means to possess the land. "A hundred years ago, our land was unknown," the children's book begins. But through the efforts of Jewish scientists, a generation of Jews in Palestine came to believe, it was now discovered.[73]

At the same time, there were others who resisted the efforts to build a scientific technocracy in Palestine, or at least registered ambivalence. Aharon David Gordon, a charismatic Russian who moved to Palestine at forty-seven in 1904 and began to work fields near the Sea of Galilee, criticized the unquestioning appreciation of science and technology that he saw in so many of his compatriots. It is easy, Gordon believed, to overestimate the value of a scientific outlook. He wrote that "it is in fact the savage, the natural man who possesses sharp, healthy, lively and intact senses. He may achieve with his simple senses what a man of science will never achieve with all of his tools and instruments."[74] Indeed, Gordon wrote elsewhere, "one of the primary causes of the defeats [referring to the first failures in agricultural experiments in Sejera and Kinneret] was that they were bound too tightly to rational methods and measures of the time and the sciences of the time."[75] It is not hard to identify strains of romanticism in Gordon's views; technology can provide, he wrote, "the flight of an airplane or a zeppelin with all their noise and nuisance, but not the flight of the eagle, nor of a pigeon, nor even of a small bird. A hymn sung out of a phonograph, but not a song sung out of a living human."[76] Still, Gordon is best seen, not as an opponent of using Western science and technology to advance Jewish settlement in Palestine (he himself supported the establishment of the Hebrew University),[77] but simply as someone more alert than most to the reverence in which science was held by his compatriots, and more ambivalent about this affection. He distinguished between "a nation that essentially lives with nature" and one "that is entirely unmoored from nature and established fully upon urban relations and the desire to conquer the land mechanically, with technology."[78]

Gordon's ambivalence was shared by some of his contemporaries, especially those who viewed themselves as acolytes of the great "old" man. In a few cases, idealists in early *kevutzot* and kibbutzim argued deep into the night about whether it was moral to purchase a power thresher or binder. The tensions between romantic and the more technocratic approaches toward developing a Jewish presence in Palestine would remain until after the state was established. But it was not long before even the most romantic of Zionist settlers came to embrace technology and the science that stood behind it, at first with ambivalence and then with enthusiasm. At the end of the late-night meetings, the answer was almost inevitably the same: Yes to the thresher. Yes to synthesized fertilizer, insecticide, and herbicide. Yes to scientifically bred grain and fruit and vegetables. It was not long, even on the most militantly ideological kibbutzim, before the romance of back-to-the-earth pioneering was grafted to the romance of laboratory-bred *yiddisher-kupfitude*, the putative Jewish genius that allowed settlers to wrench from arid Palestine better yields than had been witnessed since Joshua and the spies observed that this was a land of milk and honey.

Science and Zionist Image and Identity

It was no coincidence that the pioneer-scientist was a hero to kids caught up in the drama of building a Jewish homeland; the fact was that science and technology fit snuggly with the way many Zionists saw themselves and wished to be seen by others in Palestine and throughout the world.

Science and technology helped establish Jewish title to the land, sometimes explicitly (as by archeologists who documented generation after generation of Jewish hold on the land, reaching in an unbroken chain back to Joshua in Jericho), and sometimes through a more complicated chain of reasoning. Science and technology made plain the notion that Jewish settlement of Palestine was, in the end, a Western project flush with Western ideals and committed to advancing those ideals in the East.

In *The Land of Promise*, Jewish-Polish director Juda Leman brought this notion to life on the screen. Doleful images of local Arabs threshing with stone tools and roughly harnessed beasts, then eating and praying, open the film. After the ancient Jews were exiled from Palestine, the narrator explains, "primitive life returned." In Jerusalem, "the streets and bazaars of the old city are the same today as they were in the Middle Ages." A song plays over the opening credits, with lyrics by Natan Alterman, perhaps the greatest poet of Jewish mandatory Palestine: "A humiliated land, a god-forsaken land, / Sand and camels, sea and malaria, / Who knows why / It is so sad and pitiful?" Presently, the film shifts to Jewish settlers, working, employing boring machines to drill for water, laying telegraph and telephone cable, using tractors to clear fields and prepare roads. "With the most modern machinery," the narrator says, "the Jews are bringing Palestine back to its long-neglected fruitfulness."[79] The film also introduced another poem by Alterman, commissioned by the director. Called "Morning Song," it contained a paean to the sort of "progress" the film worshipped above all: "We love you, our Homeland . . . / We will dress you in a gown of concrete and cement."[80]

As the film toured Europe and America to great acclaim (winning, for instance, the Jury Prize in the Venice Film Festival), this message did not escape notice. "First, the film presents specimens of stubbornly non-progressive Arabs, living in medieval filth and ignorance, unaware of the potential riches in the arid soil beneath their feet," wrote a reviewer in the *Baltimore Sun*: "Then, in vivid contrast, one sees how the returning Jews, with up-to-date engineering methods and means of production, have transformed barren lands into fertile vineyards, olive and orange groves and fields of abundant grain. Next the infant industries are shown. . . . There are glimpses of such interesting institutions as the national bank operated for public and not private profit, the Hebrew University and its symphony orchestra, . . . and the lively modern city."[81]

Much the same message was the point of the Jewish Palestine Pavilion for the 1939 New York World's Fair. The plan, one of the organizers wrote, was to emphasize "the transformation of the country by modern intelligence, the use of the best modern technical resources, courage, self-reliance, faith and hard work." The pavilion's six galleries—the Hall of Agriculture and Resettlement, the Hall of Town Planning and Communications, the Hall of Industry, the Hall of Culture and Education, the Hall of Health, and the Hall of Labor and New Social Forms—highlighted the technological savvy of the settlers. A vast mosquito greeted visitors to the Hall of Health, symbolizing the success of Jewish doctors in eradicating malaria. The handbook to the pavilion encouraged guests to view a large statue standing at the foreground of the Hall of Industry: "This statue . . . is of Lot's wife. You remember that she and her husband ran away from Sodom and Gomorrah, which God destroyed and covered with the waters of the Dead Sea. She was told not to look back—but she did, and was changed to a pillar of salt. Today, she symbolizes our determination not to look backwards."[82]

It was not just on foreign shores that Jewish Palestine sought to present itself in such forward-gazing terms. Beginning in 1923 Tel Aviv put on fairs and exhibitions at irregular intervals but often.[83] The first exhibition was a modest affair, occupying three and a half rooms, showing to their best effect products of local industry and agriculture. Thousands showed up to see. In the following year, 1924, a larger exhibition with more and shinier products was held, and in 1925 came one larger still. International exhibitors joined in, bringing the newest technologies from the West to the Levant. In 1926, there were two fairs, perhaps one too many for a city of only 40,000 residents. The pace slowed, but the ambitions of the fairs grew in 1929, 1932, 1934, and 1936 (these last two held in a huge fairground built on the southern bank of the Yarkon River, then the distant north of the city); after this, local and international political upheavals initiated a hiatus that would last until 1962.[84] These last fairs attracted hundreds of thousands of visitors, from dozens of countries, swelling for their brief efflorescence the population of Tel Aviv to record after new record.

The Palestine Near East Exhibition and Fair, as it came to be called, was at heart a trade fair, sporting pavilions by local businesses made good and by foreign vendors seeking to sell their products (automobiles, factory automation, farm machines, movie projectors, as well as mundane consumer goods) to local worthies, banks, labor unions, bus cooperatives, and the Tel Aviv municipality. By the forward-looking standards of the New York World's Fair, the Near East Fair of Tel Aviv was planted firmly in the here-and-now. Yet, in its own way, it was very much about the rapid march forward of the Jewish settlers in Palestine, owing in part to what Tel Aviv mayor Meir Dizengoff characterized as the "heart, energy and enormous drive" of Jewish businessmen and laborers, but also resulting from their quick embrace of the newest, smartest, and best scientific techniques and technologies.[85]

A photo of British high commissioner Sir Herbert Samuel's visit to the 1924 Levinstein and Shulman Pavilion shows a stately entourage posed before a poster and banner bearing the name of the company's house brand—Progress chocolates, candies, and halvah. (Businesses took up the cachet of the term "progress" to brand numerous products. Shoppers in Tel Aviv in those early fair years could buy Progress furniture, Progress laundry soap, and Progress carbonated beverages; they could avail themselves of Progress laundry and dying services, Progress custom machining, or Progress driving school; and they could learn English by correspondence through the Progress Institute.)[86] In 1925, Levi Rabinovitz, a Tel Aviv electrician, advertised to exhibitors his skill in producing "especially large sign[s] of thousands of electric bulbs for the purpose of moving and changing advertisement."[87]

In 1926, fair organizers added under-the-stars cinema, showing local films of heroic settlement activities under way far from the city.[88] The Jewish National Fund imported a gramophone, still a scarce technology in Palestine of the day, and played Hebrew, Russian, and Italian speeches of celebrated Zionists like Ze'ev Jabotinsky, Nahum Sokolov, and Arthur Ruppin, to the wonderment of the large crowds that gathered.[89] Jabotinsky repaid the favor when, in opening the 1929 fair, he said: "We applaud the organizers of this fair. . . . These people were among the first in the Land who believed in the future of the factory, at a time when almost everyone who thought himself 'serious' mocked the hope that industry could develop in our land of 'pure agriculture.' They wrote 'Industry' on their standard."[90]

In fact, the fair's organizers went to some lengths to emphasize both industry and agriculture and, for good measure, Zionist government, with its initiatives in public health, economic development, cooperative settlement, and infrastructure creation. Common to all was an emphasis on scientific organization and the swift adoption of technological innovation. Pride that Jews were bringing new energy and modern ideas and artifacts to Palestine settled over almost everything at the fairs, like the fine sand that quickly accumulated on pavilions baking in the Mediterranean sun. It was evident in the triumphant description in *Ha'aretz* of the visit of British high commissioner John Chancellor to the 1929 fair: "The Commissioner and his entourage passed through several of the pavilions that they had not visited before the opening ceremonies. From the Main Pavilion, in which the Products of Palestine are concentrated, the Commissioner went to the Municipality [of Tel Aviv] Pavilion, and from there to the great pavilion of the automobiles. This pavilion is organized in a very lovely way, and the automobiles it contains are of the most exalted sort. When the Commissioner visited, the lights [of the cars] were ablaze in front and back."[91] Automobiles joined demonstrations of the newest telephones, household appliances, sewing machines, and other symbols of Jewish Palestine's adoption of the best of the West.

There was a double claim implicit in this celebration of the embrace of advanced tools and methods. Science (and scientific management and organization) and technology (and scientific industry) made plain the *progressive* nature of the Zionist undertaking, which endorsed, as Weizmann put it, "the pursuit of every form of knowledge which the mind of man embraces" and sought "a commonwealth of learning freely open to all men and women of every creed and race." (The great Gottingen mathematician, Edmund Landau, taking part in laying the cornerstone of the Einstein Institute, emphasized that "pure science knows no borders between nations, and who will give that this view will penetrate the hearts of those yet far from it?")[92]

At the same time, science and technology were at the heart of arguments that Jewish settlement of the Holy Land would benefit even those ostensibly "primitive" locals who might be displaced by the Zionists, by bringing them culture of universal value and by providing a bridge between these backward societies and the advanced West.[93] This was not lost on the locals themselves. A reporter from *Al-Jazeera*, a Jaffa newspaper, reported that he had visited the 1926 "Jews' fair in Tel Aviv" and left deflated, having seen how the land's minority was proudly presenting its progress while the Arab majority failed to press itself forward. Supporting the Jewish fair, he felt, would only make matters worse. "We must not enter under their flag and present our works in their fair," he advised.[94]

The notion that Jews could bring modern civilization to the Levant was an old one that found expression in the earliest Zionist writings, as when Moses Hess quoted Ernest Laharanne at length in his own proto-Zionist masterpiece, *Rome and Jerusalem*: "A great calling is reserved for the Jews: to be a living channel of communication between three continents. You shall be the bearers of civilization to peoples who are still inexperienced and their teachers in the European sciences, to which your race has contributed so much."[95]

Bringing European sciences to local inhabitants was assumed to be a benefit that more than justified the hardships that accompanied the arrival of the Zionists. The impact of the wise use of science and technology provided further justification. Zionists pointed to increased agricultural production —what they often called "making the desert bloom"—as a sign of the righteousness of their endeavor. This attitude, which blended a belief that success and increased efficiency are self-justifying with a belief that improvements of the land through labor goes part of the way toward conferring ownership, was rarely analyzed critically, but it had persuasive force (some of which it retains, in Israel, to this day). It was an attitude that was vigorously colonialist, stitching seamlessly a sense of sure intellectual superiority, inerrant entitlement, and selfless virtue. Through superior Western science and technology, Jews would win Palestine, and from this conquest no one would benefit more than the indigenous Palestinians. It is no wonder that this formula had such urgent appeal.

Science also served to link the promise of Zionism with the achievements of generations of Jewish scientists abroad and their famed genius.[96] It was no coincidence, nor was it idle generosity, that led the Hebrew University to name its Institute for Physics and Mathematics after Einstein (just as it had been no coincidence, as Einstein himself observed, that he was pressed into service schnorring for Zionism in "Dollar-land").

One of the great appeals of having an Institute for Physics and Mathematics in the first place, for some Zionists in any case, was the link it suggested between what was being done in the dusty alleys of Jerusalem and what Einstein had achieved in Bern twenty-odd years earlier. Zionist leaders like Chaim Weizmann, when they stumped for support, were careful to remind their audiences of the German, American, and Russian Jews accumulating Nobel prizes, promising that these numbers—miraculous though they were—would be dwarfed by the achievements of Jewish scientists working under the flag of a Jewish state.

And just as science served to associate Zionists with the achievements of refined and educated Jews in Europe, it served as well to dissociate Zionists from other, more religious Jews they had left behind. This was of great importance to some Zionist theorists. Samuel Joseph Ish-Horowitz wrote that "the Jew must negate his Judaism before he can be redeemed."[97] Judaism, defined as traditional observance, was seen as standing in the way of human redemption. Marcus Ehrenpreis was less charitable still to traditional Judaism:

> We have liberated ourselves from the shackles of a sickly, rotten, and
> dying tradition! A tradition that cannot live and does not want to die;
> a tradition that manacled our hands, blinded our eyes, and confounded
> our hearts, that darkened our heavens and banished light and beauty and
> tenderness and pleasantness from our lives, that turned our youth into old
> men and our elders into shadows. We have liberated ourselves from the
> excessive spirituality of the Exile. . . . We have liberated ourselves from the
> rabbinic culture, which confined us in a cage of laws and restrictions.[98]

Such displays of rationalist scorn were commonplace in Zionist circles. A leitmotif of much Zionist prose is a yearning for *normalcy*, for a culture in which Jews were not merely, or even particularly, Jews, but rather human beings. As the speeches at the founding of the Hebrew University made clear, nothing was viewed as more human, more universal, than the pursuit of science.

There was another side to this as well. For some early Zionists, eugenics (what its founding father, Frances Galton, called "the science which deals with all influences that improve the inborn qualities of a race; also with those that develop them to the utmost advantage") promised scientifically to improve the stock of Jews.[99] Eugenics thrived in Europe at the moment that a good number of Zionist ideologues and activists came of age, and it seemed to some an indispensable tool for those wishing not only to build a homeland for Jews, but to reform Jews as befits a new homeland. Mordechai Bruchov, then the physician of the Herzlia *gymnasium*, or high school (he later directed the Department of Hygiene of Hadassah Hospital), wrote in 1922 that "the prevailing spirit is the idea that the greatest sin that people can commit to the god of life is to procreate sick children. . . . In the struggle of nations, in the clandestine 'cultural' struggle of one nation with another, the one wins who provides for the improvement of the race, to the benefit of the biological value of the progeny."[100]

Scholars disagree about the extent of the purchase eugenic thought had among Jewish intellectuals in Palestine. In some circles, its appeal was strong, and it was dual. Embracing a "scientific" attitude toward population growth and demographics, the Zionists were once again demonstrating their willing ability to use up-to-date science to advance their social and political goals. In a more practical vein, for those so persuaded, eugenics was a science that could not just improve the material conditions of Palestine's Jews, but also actually improve the Jews themselves.

By embracing science, then, some Zionists deliberately associated the Zionist project with the progressive West and with the great achievements of generations of Jewish scientists abroad, while dissociating it from the primitive, overly *Jewish* Jews of the shtetl. Along the way, it provided ample justification for the colonization of what most Zionists saw as hopelessly primitive Palestine. Jews would bring the best of the West to the Levant— wealth, culture, comfort—and they would do it using those tools that Jewish hands had wielded so capably in the West: the tools of modern, universal science. In these ways, science served perfectly the *ideological* agenda of political Zionism in the decades before Israel was established.

Science, Technology, and State-Building

But this was not all there was to it. Science (and technology) increasingly met the growing practical needs of Jews in building a national infrastructure in Palestine. When Lord Peel, the chairman of the Palestine Royal Commission, was dispatched from England in 1936 to consider limiting further Jewish immigration to Palestine and found Weizmann tending test tubes in his Rehovot laboratory, he inquired what the scientist was doing. Weizmann replied, "I am creating absorptive capacity."[101] Behind this jest was Weizmann's dead serious belief that only through science would it be possible to feed, clothe, heal, and protect Jews in Palestine as their numbers grew. This was a matter of surpassing importance to Zionists. Already in 1891, Ahad Ha'am (Asher Zvi Hirsch Ginsburg, the founder and champion of "cultural zionism") had warned in an essay called "The Plain Facts about the Land of Israel": "Those who extol the land have conceded to their opponents that their homeland can not presently receive the multitudes of our people who are wandering from their native land. This applies particularly to merchants and artisans who are seeking an immediate source of livelihood and are incapable of preparing all that is required for agriculture and waiting for the land to give forth its fruit."[102]

The desire to overcome this problem was a source of enduring anxiety for Zionist leaders and a continuous spur to deploy those varied sciences—from agronomy, chemistry, mineralogy, geology and biology to Taylorist efficiency science—to increase efficiency and production, and thereby increase Palestine's carrying capacity for Jews.[103] This too was part of the reason that, for Weizmann and many others like him, the ideological and the practical appeal of science were inseparable. The degree to which this was so became obvious in the decade immediately before and the decade immediately after the establishment of Israel in 1948.

The first years of the country's existence established Israel as a technocracy willing to devote great resources to developing science and technology, its leaders certain that economic, political, and social success *depended* on science and technology. It was during this time that the Israel Defense Forces (IDF) formulated its basic strategy of maintaining technological and scientific superiority at all costs and founded the laboratories, R&D facilities, and factories necessary to the implementation of this strategy. David Ben Gurion, the chairman of the Jewish Agency and in time the powerful first prime minister of Israel, had outlined this strategy in a 1947 memo he sent to the head of the Zionist military organization, the Haganah, setting forth a plan for winning by force and maintaining a state: "Our human materiel in general is . . . immeasurably better in its moral and intellectual ability than our neighbors. This is our main advantage and at the moment almost our only advantage . . . We must . . . [take] advantage of all the achievements of state-of-the-art science and technology for our defensive needs."[104] Ten years later, much in the spirit of Ben Gurion's memo, and with his enthusiastic support, Israeli physicists were hard at work building an atom bomb.[105]

It was during these decades that the first huge technological development projects were dreamed up and carried out, and it was during these years that the allure of such huge projects grew so great that no one thought to question them. The Hula Swamp was drained with a passion and brio so heartfelt that, from today's remove, at which this massive feat of engineering is generally seen as an ecological tragedy, one's heart breaks to conjure the spiritual devotion with which the project was carried out.[106] Plans for desalinization plants, for satellites, for alternative energy, were hatched with like enthusiasm.[107] These big engineering projects inspired national pride, and many took them as the fulfillment of an ideal that had accompanied Jewish settlement in Palestine almost since its first days.

Draining swamps, farming deserts, desalinating water, launching satellites —all these helped make Israel in its first years a model for science- and technology-based development for emerging nations. It was during this time that science and technology transfer first became an Israeli diplomatic stock-in-trade, especially in the Third World. In 1953, Foreign Minister Moshe Sharett agreed to send agricultural and aeronautics experts to Burma, a relationship that blossomed into diplomatic relations and led, in 1955, to a state visit by Burmese Prime Minister U Nu, the first such visit to Israel by an Asian leader. After Golda Meir assumed the post of foreign minister, which she held from 1956 to 1966, she established scores of technical, scientific, medical, and agricultural assistance programs in Africa, a good number of which paved the way to full-scale diplomatic relations.[108] What characterized these diplomatic efforts, one scholar later concluded, was their "almost total focus on technical assistance."[109] Science and technology broke barriers where politics and traditional diplomacy could not.[110] It was this that made the 1960 Rehovot conference on science and the new states both possible and, for Israeli leaders, necessary. If the Jewish state was to be, as three generations of Zionist leaders had hoped, a bridge between the West and the East, then its paving stones were hewn from the science and technology at which Jews in Palestine and beyond excelled.

Conclusion: The Enchantments of Science and Technology

For all these reasons, it was also during this time that, as a U.S. embassy analyst put it in his brief to Washington, Israelis became "indissolubly tied to science and technology as a principal motivating factor in social and economic progress."[111]

Almost from the moment the first Zionist settlers found their way to Palestine late in the nineteenth century, science and technology had been revered for many reasons, as we have seen. Jews in Europe and America had enjoyed prodigious success in science, and with this success came the status that all Jews, including Zionists, wished to appropriate. Scientific achievement was also synonymous with progress, enlightenment, and rationality, traits that many Zionists, who were weary of being viewed in Europe as primitives, greatly admired. And just as the great colonial powers—England and France first among them—used science and technology as a way to justify their occupation of far-flung lands, arguing that they were bringing progress and modernity to their backward wards, so too did Zionists, in the cheery confidence that they were saving the Levant, bringing it into the twentieth century rather than appropriating it.[112] It was also true, as Abba Eban and the other organizers of the 1960 Rehovot conference realized, that the young state's accumulating scientific and technological achievements played a role in Israel's acceptance in the world community as a "Western" country, in defiance of its geography, and allowed it to serve as a conduit for transmitting Western techniques and values to the developing world. No less, the confidence of earlier Zionists like Chaim Weizmann that science and technology would be the key to solving the raft of problems faced by the Jewish state—from feeding and clothing waves of immigrants, to protecting the country from outnumbering neighbors who might turn belligerent, to multiplying the subsistence standard of living they found when they arrived in Palestine to levels that would keep Jews from streaming to the United States and other wealthier destinations—has been handsomely confirmed by a century of remarkable achievements in development and defense, owing to a century of remarkable achievements in science and technology. In all these ways, science and technology played a role from the very start in the way many Zionists understood who they were and what they were doing. Equally, science and technology played a role from the start in who the Zionists became and what they did.

Precisely what forms the "indissoluble" ties linking Israelis over the past fifty years to science and technology have taken is a question of exquisite complexity that scholars have only recently begun to tackle, and as of yet, only partially. Still, for even the casual observer, links of one sort or another are everywhere to be seen. After many years of fretful absence from the list of Nobelists, the past decade has brought Israeli scientists six Nobel Prizes, to great fanfare. They've found a place as national heroes alongside entrepreneur-millionaires made rich by high-tech innovation, and their advice has been sought on television and radio about everything from politics to fashion. A popular book describes Israel as a "start-up nation," a designation one of its authors explains like this:

> Israel has the highest density of tech start-ups in the world. More importantly, these start-ups attract more venture capital dollars per person than any country—2.5 times the U.S., 30 times Europe, 80 times India, and 300 times China. Israel has more companies on the tech-oriented NASDAQ than any country outside the U.S., more than all of Europe, Japan, Korea, India, and China combined. But it's not just about start-ups. Scratch almost any major tech company—Intel, Microsoft, Google, Cisco, Motorola, and so on—and you will find that Israeli talent and technology play a major role in keeping these multinational companies on the cutting edge.[113]

Science-driven high tech is also now at the heart of Israel's army and accounts for much of whatever day-to-day security many Israelis feel. On

September 3, 2012, first place in the Israel Security Prize, the Army's Oscars, was awarded to the developers of the "Iron Dome," an automated, ground-to-ground antimissile system developed in just three years, a span so short it is without precedent in peacetime. President Shimon Peres, who presided over the award, thanked the engineers for their genius and creativity, adding: "Thousands of citizens today feel safer because of the Dome. They owe their security to batteries [of missiles] that destroy missiles while they are in the air and prevent many injuries, especially in settlements near Gaza from which many missiles are fired."[114]

Science and technology are today a powerful part of the self-image of a great many Israelis, and a source of considerable comfort. Science and technology allow Israelis to sleep at night. Science and technology allow Israelis to feel superior to their hostile neighbors. Science and technology promise a wealthier and healthier future. Science and technology also seem to link Israel with the world—as a universal endeavor—but also with its particular Jewish heritage, as a token of Israel's perpetuation of putative Jewish scientific genius.

In Israel's national amphitheater in Rehovot, the following quotation is carved on stone tablets: "I feel sure that science will bring to this land both peace and a renewal of its youth, creating here, the springs of a new material and spiritual life. And here I speak of science for its own sake, and applied science."[115] The words are Chaim Weizmann's; the amphitheater was built beside his hilltop garden gravesite. On cool spring days, schoolchildren are herded to see the great man's final resting place, and the words may be no less true for them than they were for Weizmann himself. As they snap photos of the grave on their iPhones, one can sense just how deeply science and technology have penetrated the material and spiritual lives of these young Israelis, just as it penetrated the lives of their parents and their parents' parents.

When All Worlds Were New Worlds

A S HISTORIAN YURI SLEZKINE WROTE, Jews at the start of the twentieth century had three great "destinations"—the metropolises of America, the great cities of the Soviet Union, and the arid rough of Palestine—each representing "alternative ways of being modern." Even before Nazis had destroyed most of Europe's Jews, these three destinations had become capitals of Jewish life. Well before mid-century, most Jews called one or the other of them home.[1]

In each of these places, as we have seen, science played a large role in the lives and livelihoods of Jews, and not only that—science played a similarly large role. In New York, Moscow, and Jerusalem, Jews extolled the virtues of science in different languages but in like terms. In all three destinations, Jews of substance praised science as a means to create a more perfect society, to better defend their adoptive homeland, and to more surely advance all of humanity. There were those, too, who said that science might help to demonstrate once and for all that Jews had finally earned a place in the society in which they lived, in their homeland, in the family of humanity.

This similar affinity for science that one finds among Jews in the United States, the USSR, and mandatory Palestine is a puzzling historical fact. What accounts for it? The three great Jewish destinations differed in large ways and small; indeed, they could hardly have been more diverse. As Slezkine observed, each had its own *ism*: liberalism in the United States, Communism in the Soviet Union, and Zionism in Palestine. The impacts of these broad ideologies on the lives of those who lived in their light and shadows were complicated and variegated, but they were never negligible. The material challenges facing Jewish immigrants in each of the three destinations differed greatly. Just as Jewish pushcarts were appearing on New York's Lower East Side, they were disappearing in Moscow's Otrendoe neighborhood, banned as an ugly vestige of an outmoded social economy—the same reasons for which they were disparaged in the Hadar quarter of Haifa. The cultural challenges facing Jewish immigrants in each destination differed greatly, too. What it meant to be a Jew, or to cease to be a Jew, was largely a function of geography, leaving Americans a set of options—Reform, Conservative, Orthodox, Reconstructionist, Ethical Culture, atheist, and more—that differed greatly from the choices faced by Soviet Jews and, in turn, by Jews in Palestine. At the same time, the institutions in which sciences were taught and practiced differed enormously from place to place.

That these three very different places produced among Jews attitudes toward science that were not *so* very different—and in some aspects were quite similar—is a curious fact that demands an explanation. It is like the scenario in a sci-fi movie when a crew from earth settles on a distant planet and discovers creatures who breathe air and speak English. At first this seems natural enough; upon reflection, it comes to seem impossibly, laughably, *weird*. It defies reason and begs explanation.

When explanations have been given in the past, the similar pull that science had for Jews in the United States, the USSR, and Palestine has usually been attributed to traits of mind putatively shared by these distant Jews, to their similar *yiddisher-kupfitude*. Sometimes these explanations take a biological slant (generations bred Jews to be clever, endowing them with naturally selected brains that excel equally under capitalism, Communism, and Zionist collectivism) and sometimes a cultural slant (because of generations of venerating learning, Jews were trained to hit the books with enthusiasm, be they Talmudic tractates or physics monographs). Whether biology or culture, these explanations posit that Jews are Jews and that something in their character accounts for the alacrity with which they take to science, whether they are in Chicago, Kharkov or Kfar Saba.

As I wrote in this book's introduction, explanations of this sort do not survive scrutiny. For one thing, they offer no account of why, with only a few exceptions, Jews were mostly indifferent to science and displayed no special talents for it until the first decades of the twentieth century. For another, the notion that natural selection endowed Jews with special genius in the short span (by evolutionary standards, anyway) that they have been around is a biological implausibility. And while it may be true that learning held a place of honor among many Jews in the generations prior to the twentieth century, it is equally true that relatively few Jews at the beginning of that century had benefited from sustained and nourishing educations, religious or secular. Many, probably most, of the Jewish PhDs and professors described in this book came from homes with few books and even fewer framed diplomas hanging on the wall. The world of letters is in desperate need of a good book about Jewish benightedness; in its absence, we overestimate time and again the bookish cultivation of Jews.

But if Jewish brains and Jewish habits of mind are not enough to explain why Jews in the three very different places described in this book shared a similar affinity to science, what is? What thread was common to the three destinations?

At the simplest level, the common thread was a *web* of common threads. Although the three destinations lay at great distances (both physical and ideological) from one another, and though transcontinental travel in the first decades of the twentieth century was exhausting and expensive, links between the communities were durable, enduring, and spirited. An urgent paradox of Jewish history in the first half of the twentieth century is that while it cannot be accurately described outside of national contexts (American Jewish history is ineluctably *American* Jewish history), at the same time it cannot be accurately described within national borders alone (American Jewish history makes no sense except in light of European Jewish history and Zionist history).

In December 1906, by way of illustration, Dr. Shmaryahu Lewin came to Manhattan from Russia after a portentous summer. In May of that year, Lewin had been the first Jew elected in the first Russian Duma, and as a representative of the Constitutional Democratic Party, he put forth proposals for expanding the civil rights of Jews, eliminating Russia's death penalty, and reducing the authority of the tsar. By July, Nicholas II had dissolved the Duma. The state of affairs in Russia and, especially, the status of the Jews who remained there was of intimate interest to the multitudes of émigrés building new lives for themselves in New York, and throughout his three-week visit to New York, Lewin was greeted by throngs eager to hear his report and assessment. Arriving at the Durland's Riding Club on Central Park West, Lewin was met by "probably the largest gathering of Jews ever seen in the city"[2] (well over ten thousand people, according to the *Times*, including "Jews from every quarter of the metropolis, poor Jews and rich Jews, laboring men and capitalists").[3] The banker and Jewish leader and philanthropist Jacob Schiff opened the evening, observing that the fate of Russia's Jews "is a cause which makes our hearts beat higher and stronger" and "one with which the people of this city are in sympathy."[4] "Give the Jew the right of citizenship," he continued, "and whereas today they call him a curse, they will then call him a blessing."[5]

Then Lewin proceeded to speak, in German until the crowd chided him to switch to Yiddish. He spoke for eighty minutes to a hushed, rapt throng about the sorry state of Russia's Jews, about the unsteady hope of the Duma, about the impressive successes of his countrymen who had moved to America, and, in conclusion, about the importance for Jews everywhere to devote themselves to building a Jewish homeland in Palestine. When he finished speaking, the crowd swelled toward the stage, buckling the rostrum and sending dozens tumbling. The *New York Tribune* reported that Golde Rabinowitz, wife of Sholem Aleichem, the acclaimed Russian Jewish writer who had immigrated to the city only a year before, "became hysterical when she thought her husband had gone down in the crash," recovering when she learned he was unharmed.[6] Lewin also spoke to rapturous crowds in Boston. Over the next years, he would return to Russia (thousands saw him off at Grand Central Plaza),[7] move to Germany, and return to America, before finally moving to Palestine. Throughout these years, and in all these places, he gave much of his energy to raising money and gathering political support for the establishment of the Technicum in Haifa.[8]

Lewin was a remarkable figure, but in being at once a figure of importance in Russia, the United States, and Palestine he was hardly unique. Chaim Weizmann, of course, was Russian, English, and Palestinian (in time, Israeli), and a figure of adulation in the United States.[9] Many other examples could be provided, but the point is a simple one: traffic between the newly emerging centers of Jewish life—in America, in the cities of Russia (and then the Soviet Union), and in mandatory Palestine—was constant and lively and filled with substance.[10] It is hardly surprising that certain cultural patterns—among them, the affinity toward science that emerged in each center—would be part of this traffic.

But deeper forces were also at play. For one thing, it was not just Jews in America who had come to "the New World." Jews leaving the Pale and settling in the cities of revolutionary Russia had come to their own new world. And Jews streaming to Palestine had come to a new world of *their* own. For most Jews in the first decades of the twentieth century, there were only new worlds. This was a fundamental commonality among a great many Jews, more fundamental than much of what divided them.

In the cases of Russia and Palestine, it was clear even to those of the most modest aspirations that momentous change could be at hand. In Russia, even in the years before the revolution—years of pogroms and of fitful, half-hearted, and ephemeral reforms—Jews had good reason to believe that their grandchildren would inhabit a society vastly different from that of their grandparents.[11] After the revolution, this could hardly be doubted.

On April 14, 1917, the Central Committee of the Russian Bund, the Jewish Communist Party in Russia, sent a cablegram from Petrograd to New York addressed to "all Jewish Workingmen of America":

> Today the Central Committee of the Bund congratulates the American proletariat and all American Jewish workmen upon the greatest victory the working classes have ever achieved. Great and invaluable is the international importance of the Russian revolution. Today Russian working-men enter the world's democracy as equals. The time of the international is nearer, the world is free from gendarmes, and the revolutionary energy of the German proletariat is awakening. Peace will be made by really free nations. . . . The Russian Revolution opens new avenues to all Russian Socialists, to Jewish workingmen and the Bund. . . . The full realization of the social democratic program . . . is no more a dream, but a real possibility. . . . With one blow the Russian Revolution has conquered Czarism, abolished all restrictions, and opened a new page in Jewish history. The liberation of the Jewish nation is in the faithful hands of the revolutionary Russian nation.[12]

At their most hortatory and hopeful, Jews streaming to Palestine and those Zionists who supported them from distant shores were hardly less confident of the revolutionary importance of their undertakings. From Mendes' prediction that a Jewish state would visit upon the world "universal peace, universal brotherhood, universal happiness,"[13] to Herzl's belief that it would provide a laboratory in which the best ideas of the West could be tried and perfected, to the faith of early kibbutz members that they were rewriting the rules of agriculture, economics, sex, family, and ultimately, human nature, to Abba Eban's assurance that a new Jewish state could do for the developing world what none of the great nations of Europe or North America could, a great many Zionists were brashly certain that theirs was a revolutionary undertaking of global scale and timeless importance. Even a shopkeeper in Tel Aviv, making a living in a way not much different from that of his father or grandfather, might well believe that he was forging a new and better world.

And the same could be true for Jews in America, that most unrevolutionary of new worlds. Bernard Gerson Richards, a journalist and the founder of the Jewish Information Bureau of New York, introduced a 1903 essay, "Zionism and Socialism," with a story: "In one of the Jewish bookstores on the East Side of New York, an ardent member of a Zionist society was offering to sell tickets to its annual ball to all who entered. A young man came in to buy his daily Yiddish paper. He was accosted by the Zionist. 'No,' said the customer, disdainfully, 'I do not want any of these tickets.' 'Ah,' said the ticket-vendor, angrily, 'you are one of those Socialists.'"[14]

Even in New York, many Jews at the start of the century dreamed of this new world or that. Having decided to quit Russia for America, before he ever boarded a boat, Abraham Cahan found his fellow emigrants in a utopian reverie: "A spirit of prophecy was here. A man on the street would suddenly begin an impassioned speech on the world-wide brotherhood which was to grow out of our American communist colonies. Our idea, he said, would spread all over the earth; there was to be one language for all humanity, and an end to all tyranny, misery and injustice. The whole world would be changed!"[15]

Most Jews who came to America sought, not to change the whole world, but rather to improve their own meager circumstances. "In America," Cahan wrote, "a 'shister' [shoemaker] could soon become a 'mister,'" and this homely aspiration tugged on the imaginations of most immigrant Jews more than anything Marx or Herzl ever wrote.[16] But even Jews whose most radical wish was to graduate from shister to Mister found that they, too, sought a new world. Jews eagerly paging through want ads sought a world in which "Christians only" did not appear. Jews dreaming that their children would study in the best universities in the land, and then in its best medical schools, sought a world without quotas. Jews wishing to send their children to the public schools sought a world in which pupils were not forced to mouth Christian prayer, or study the New Testament, or sing of Christ's glory at Christmastime. Jews wishing to pursue economic opportunities wherever they saw them sought a new world in which the Klan had no power and no place.

For most Jews in the first decades of the twentieth century—whether they found themselves in Russia, America, Palestine, or Poland, Germany, England, France or any of the dozens of other countries which they occupied —there were *only* new worlds. The social forms that defined the lives of their grandparents were unavailable (and in any case hardly appealing) to a great many Jews in a great many places at this time.[17] Relatively few Jews even lived in the same region that their grandparents had. And as the twentieth century proceeded, fewer and fewer shared a mother tongue with their grandparents. Especially in the three great destinations that have been the focus of our attention, Jews faced new opportunities that could hardly have been imagined in prior generations, of living in new ways and forging new relationships with the rest of the world, their neighbors near and far. In the first decades of the twentieth century, this experience—of facing a future in a new world, perhaps with enthusiasm, perhaps with trepidation—was a great shared experience, perhaps *the* great shared experience, of many Jews in widely different places.

And it was this great shared experience that begins to explain the affinity and ardor these far-spread Jews shared for science. Like many others at this time, a good many believed that the rise of science was corrosive of the "old world"—the old social orders in which their prospects were limited. And like many others of the day, a good many believed that the ideals and values that science might deliver might produce societies that were at once better *and* better for Jews. Like the Scottish-Jewish mathematician Hyman Levy, many believed that science demands that statements of fact (and, by extension, whatever policies might be based on them) be "invariant with respect to the individual,"[18] meaning that science was at odds with the sort of discrimination that often stood between Jews and the education they desired, the jobs they sought, and the public standing they longed for.

With the American Jewish sociologist Robert Merton, these Jews believed that science demands honesty, integrity, skepticism, disinterestedness, impersonality, and other values that could only make the societies in which they lived more welcoming to Jews.[19] Like the German-American Jewish anthropologist Franz Boas, they were certain that science insisted that "a citizen be judged solely by the readiness with which he fits himself into the social structure and by the value of his contributions to the country's development," and not by accidents of birth and belief.[20] With the American Jewish physicist Robert Oppenheimer, many maintained that "science has brought all the different quarters of the globe so close together that it is impossible to isolate them one from another . . . [and] if civilization is to survive, we must cultivate the science of human relationships—the ability of all peoples, of all kinds, to live together and work together in the same world at peace."[21] Like Russian Jewish sinologist Vitaly Rubin, many noticed that in the scientific faculties of universities (like Moscow University, where Rubin studied), "the Jewish Question did not arise there. Not only did it not arise in the form of anti-semitism; it did not arise at all."[22] And like the Russian-English Palestinian-Jewish chemist Chaim Weizmann, many Jews saw in science "a commonwealth of learning freely open to all men and women of every creed and race [through which] political strife and division cease and all creeds and races will . . . be united in the great common task of searching for truth, [producing] a future pregnant with possibilities not only for the Jews or for Palestine, but also for the awakening East and for mankind at large."[23] And like the Polish-Russian, Turkish-educated Palestinian-Jewish David Ben Gurion, who served as Israel's first prime minister, many Jews believed that the Jewish state reflected "the best of the Western scientific tradition—scientific honesty, respect for merit, self-criticism, and reward based on competence."[24]

These very different people from very different circumstances with very different aspirations understood the ethos of twentieth-century science in a startlingly similar way.[25] They all saw science as a progressive force that would replace arbitrary old orders (with their fealties to class, rank, and religious and cultural pedigrees that had long ensured the exclusion of all but a few Jews from the seats of power, wealth, and public esteem) with new orders based on fact not faith, achievement not pedigree, and innovation not whorish fealty to hoary ways. Also, they saw science as a universalist and *universalizing* force. ("Science is wholly independent of national boundaries and races and creeds," wrote Franz Boas in his manifesto, signed by 1,284 of the world's leading scientists.)[26]

It was the hearty appeal of this universalism for Jews in the first half of the twentieth century that seemed to Jean Paul Sartre the very soul of the Jews he had observed in Europe, Palestine, and North America. Writing in 1944, a fraught moment soon after the liberation of Paris, Sartre explained this appeal:

> Of all things in the world, reason is the most widely shared; it belongs to everybody and to nobody; it is the same to all. If reason exists, then there is no French truth or German truth; there is no Negro truth or Jewish truth. There is only one Truth, and he is best who wins it. In the face of universal and eternal laws, man himself is universal. There are no more Jews or Poles; there are men who live in Poland, others who are designated as "of Jewish faith" on their family papers, and agreement is always possible among them as soon as discussion bears on the universal. . . . The best way to feel oneself no longer a Jew is to reason, for reasoning is valid for all and can be retraced by all. There is not a Jewish way of mathematics; the Jewish mathematician becomes a universal man when he reasons.[27]

Science itself, reason itself, seemed to demand that the Jewish mathematician not only view himself as "a universal man," but that he be viewed that way by all others who took science and reason seriously. Sartre was savaged by critics of his day for missing the complexity of Jewish experience,[28] or for misconstruing Jews altogether ("I was surprised," wrote Octavia Paz, "by his saying that Judaism, the least universal of the three monotheisms, is the origin and foundation of this hope [for universal and universalist goodwill]: Judaism is a closed fraternity").[29] Sartre had, in fact, essentialized Jewish experience in a way that overlooked its great variety in different epochs and locales. But he had captured something true about his moment, when many Jews throughout the West sought to be treated where they found themselves as *human beings* rather than Jews, and saw in science an aid to reach that end. And like Mosei Gran and the other editors of the Leningrad journal *Problems of the Biology and Pathology of Jews*, they saw science as doing this in two ways at once: reforming Jews and reforming the societies in which they lived, leaving each less closed and less fraternal than they had been before.

This yearning of many Jews for a new world, one in which they could find a place on terms unlike any their parents and grandparents had known, was what was common to Jews in the very different new worlds of New York, Moscow, and Tel Aviv. It was this yearning for new worlds, too, that explains what Jewish affinity for science shares with other Jewish predilections of the day. Scholars have argued that the desire to portray on screens across America a society that embraced all its sons and daughters was a part of what drew Jews to Hollywood in such numbers, at the same time and in the same way they were drawn disproportionately to Harvard. Others have argued that a wish to help fashion a new society of Soviets, in which the old tsarish prejudices were remembered as obscene and bygone history, was part of what drew Jews to the Soviet police forces (and especially the secret police) and army in such large numbers, at the same time and in the same way they were drawn to research laboratories.[30] And it may be that a wish to help fashion a new society of collectivist equality and ruddy agrarian health was part of what drew Jews to *kvutzot* and *kibbutzim*, at the same time and in the same way that they were drawn to a Hebrew University. Science was not the only way in which Jews in the first half of the twentieth century sought to build a new world. It was not the most important. But in every place that Jews set out to reform themselves and the places they lived, a great many saw in science a tool, powerful and true, to do so. This was a source of its appeal not just to those who saw in science a vocation, but also to the Jewish lay leaders and rabbis, politicians, philanthropists and public intellectuals, writers and businessmen, and so many more whose enthusiasm for science was, in place after place, unmatched by any of the other groups among whom they lived.

Over the past half-century, this enthusiasm has lost some of its force for Jews in at least some of the places they find themselves. Throughout the West, science has lost some of its appeal over the last six decades. Hiroshima was, for many, the start of a long nightmare in which science produces menace, not progress. As deserts spread, and storms grow more violent and unruly, many wonder if science and scientific technologies have not given

rise to problems that scientists and technologists are unable to solve. When tyrants and zealots can produce deadly viruses in dirty laboratories using recipes they download from the Internet, the notion that science and scientific technologies by their nature spur progress, award merit, and militate against age-old hatreds like anti-Semitism has come to seem quaint and implausible.

What's more, the role that science plays in Western societies, and the image of the scientist, are very different now than they were when, say, Einstein wrote to Roosevelt to propose that the president take "quick action" to build an atomic bomb.[31] Einstein could advise politicians about political decisions precisely because his scientific expertise was taken to transcend politics. Science conferred on Einstein a voice not as a citizen, and certainly not as a Jew, but as a faithful representative of scientific knowledge itself. This was part of the appeal of science in Einstein's day—its ability to raise political debate above the province of gentlemen, where it had resided for so long, often to the exclusion of Jews. But science no longer has that ability, surely not to the degree that it did. "It is not usually possible any longer to depoliticize or depersonalize political decisions and actions," political theorist Yaron Ezrahi explained in a recent lecture at Harvard. "This was pervasive before."[32] Science and scientists no longer have the status they did a century ago, and they are more often seen as embattled participants in political debates (on climate change, say, or genetic engineering or vaccination or teaching evolution in public schools, etc.), rather than objective arbiters of these debates. Not only did science cease to be seen, by many, as an admirable agent of democratic values; it increasingly came to be seen, by some, as an enemy of these same values. "Believing that they were forced to choose between democratic values and the benefits of science," Andrew Jewett concluded, "many Americans were prepared to reject the dream of scientific democrats and their Enlightenment-inspired vision of a society modeled on the intellectual freedom of scientists."[33]

Such considerations are only part of the story, of course. In the United States, science provided a solution to a problem that Jews no longer face in an acute way. American Jews today, as a group, enjoy wealth that American Jews of a century ago could hardly have imagined. The same is true for access to power wherever it resides, in politics, business, law, media, and academia. If science was seen in the first half of the twentieth century as a tool for a dual reform—making Jews more fit for American society and making American society more fit for Jews—such reforms are no longer needed. Intellectual traditions decline slowly, and the strength of American Jewish opposition to teaching "intelligent design" in public schools suggests that Jews continue to hold science in higher regard than any other ethnic or cultural group in the United States.[34] Still, the high-water mark of Jewish avidity for science has passed and much of the passion has waned, as the steadily declining numbers of American Jews entering the sciences demonstrate.

The collapse of the Soviet Union and the exodus of the greater part of the Jewish communities there ended with finality the remarkable, short-lived story of Soviet Jews and sciences. Only in Israel does it live on, although even there, each year fewer students study advanced physics, chemistry, biology, or mathematics in high school. Aaron Ciechanover, the Israeli biologist who won the 2004 Nobel Prize in chemistry, rages against the decline of science he sees around him, as demonstrated by "the school students' declining results, the brain drain and the shortage of means for buying new research equipment. In much more difficult times for the state we saw David Ben-Gurion's glowing face when he dedicated the new building for the Technion's chemistry faculty. Today, not even a deputy minister from a marginal party would deign to attend such a ceremony. . . . We had a state with narrow roads and broad universities, not one with broad roads and narrow universities."[35]

Still, for all that the attraction of science has diminished, the notion that science has something of particular merit, of moral mettle, remains strong in Israel, as it does among many Jews in America and elsewhere. The romance of science has quieted as the circumstances of Jews have changed, but it has not disappeared. All one need to do to see that this is true is to take a trip with a vanload of rabbinical students to Kentucky's Creation Science Museum. Wandering the vitrines with the next generation of American Jewish leaders, registering their surprise, measuring their disappointment and their mounting despair, you will see how alive science remains in the political, social, and yes, religious imaginations of many American, and not just American, Jews. Indeed, the tumultuous last century of Jewish history, which saw both a methodical campaign to blot out Europe's Jews and the inexorable absorption of Western Jews into the societies in which they now live, makes little sense except in light of the place of honor science has occupied—for a brief and momentous time—in Jewish imaginations. Through the generations when Jews entered Western societies in great numbers, science was a vivid feature of their dreams of refashioning these societies to ensure that they maintained a place for Jews.

Preface

1. This observation was shared by the president of the Kentucky Paleontological Society, Daniel Phelps, who on the website of the National Center for Science Education called the Creation Museum an "Anti-Museum." See http://ncse.com/creationism/general/anti-museum-overview-review-answers-genesis-creation-museum (accessed Mar. 21, 2013).

2. The petition and its signatories are available at the website of the National Center for Science Education, http://ncse.com/taking-action/aig-creation-museum (accessed Nov. 1, 2012).

3. *Friends of God: A Road Trip with Alexandra Pelosi*, HBO Documentary, 2011, www.youtube.com/watch?v=Ym-HQ2EYKkQ&feature=youtube_gdata_player.

4. Ken Ham, ed., *War of the Worldviews: Powerful Answers for an "Evolutionized" Culture* (Hebron, KY: Answers in Genesis, 2006).

5. An image of the legend can be seen at http://jabinns.blogspot.co.il/2012/09/day-3-creation-museum_6.html (accessed Mar. 26, 2013).

Introduction. "Ridiculously Disproportionate"?

1. Robert Sommer, "Thorstein Veblen and Dementia Praecox," *Journal of the History of the Behavioral Sciences* 43, no. 3 (2007): 308–9.

2. Florence Veblen, "Thorstein Veblen: Reminiscences of His Brother Orson," *Social Forces* 10, no. 2 (Dec. 1, 1931): 194; Sommer, "Thorstein Veblen and Dementia Praecox," 309. In 1919, Veblen accepted an appointment at the newly formed New School for Social Research, returning yet again to the academy, which he regarded with divided affection.

3. Thorstein Veblen, *The Higher Learning in America* (New York: B. W. Huebsch, 1918), 42.

4. Thorstein Veblen, *The Engineers and the Price System* (New York: B. W. Huebsch, 1921), 94–97.

5. Thorstein Veblen, "The Passing of National Frontiers," *The Dial: A Semi-Monthly Journal of Literary Criticism, Discussion, and Information (1880–1929)*, Apr. 25, 1918, 390.

6. Thorstein Veblen, "Bolshevism Is a Menace—to Whom?," *The Dial*, Feb. 22, 1919, 174–79.

7. Thorstein Veblen, "The Intellectual Pre-eminence of Jews in Modern Europe," *Political Science Quarterly* 34 (1919): 33–37.

8. Veblen, "The Intellectual Pre-eminence of Jews," 34–35.

9. Veblen, "The Intellectual Pre-eminence of Jews," 39.

10. A brilliant reevaluation of Veblen's thesis, appreciative and critical at once, can be found in David A. Hollinger, "Why Are Jews Preeminent in Science and Scholarship? The Veblen Thesis Reconsidered," *Aleph*, no. 2 (Jan. 1, 2002): 145–63.

11. Lawrence Van Gelder, "C. P. Snow Says Jews' Success Could Be Genetic Superiority," *New York Times*, Apr. 1, 1969, 37.

12. This reasoning proceeds from start to finish by leaps. As Sander Gilman has suggested, the link between the two clauses in the compound sentence, "Jews are smarter than the average non-Jew and they are predisposed to become scientists" is hardly one of logical entailment. And, of course, each clause itself can reasonably be called into doubt. See Sander L. Gilman, "'The Bell Curve,' Intelligence, and Virtuous Jews," *Discourse* 19, no. 1 (Oct. 1, 1996): 65.

13. Reuben Hersh, "Under-represented Then Over-Represented: A Memoir of Jews in American Mathematics," *College Mathematics Journal* 41, no. 1 (Jan. 1, 2010): 5–6.

14. Norbert Wiener, *Ex-prodigy: My Childhood and Youth* (New York: Simon and Schuster, 1953), 11–12.

15. Lewis Feuer, "The Sociobiological Theory of Jewish Intellectual Achievement: A Sociological Critique," in *Ethnicity, Identity, and History: Essays in Memory of Werner J. Cahnman* (New Brunswick, NJ: Transaction Books, 1983), 93.

16. Nathaniel Weyl, *The Creative Elite in America* (Washington, DC: Public Affairs Press, 1966), 2.

17. Nicholas Wade, "Researchers Say Intelligence and Diseases May Be Linked in Ashkenazic Genes," *New York Times*, June 3, 2005, science sec., 1; Gregory Cochran, Jason Hardy, and Henry Harpending, "Natural History of Ashkenazi Intelligence," *Journal of Biosocial Science* (2005): 1–35. For a broader analysis of the place of "biological discourse" in the identity of American Jews, see Shelly Tenenbaum and Lynn Davidman, "It's in My Genes: Biological Discourse and Essentialist Views of Identity among Contemporary American Jews," *Sociological Quarterly* 48, no. 3 (July 1, 2007): 435–50.

18. For a recent discussion of the question of whether Ashkenazi Jews are genetically predisposed to scoring high on IQ tests, see Richard E. Nisbett, *Intelligence and How to Get It: Why Schools and Cultures Count* (New York: Norton, 2010), 173–78. Nisbett concludes that "there are many genetic theories of Jewish intelligence but not much in the way of convincing evidence." Polymath scholar and critic Sander Gilman devoted a book to explaining why such theories proliferate despite the absence of convincing evidence. As for the question, "Are 'the Jews' smarter than everyone else?" Gilman's answer is exquisitely subtle: "They are smarter and different only if the cultures in which they dwell need them to

be smarter and different." Sander L. Gilman, *Smart Jews: The Construction of the Image of Jewish Superior Intelligence* (Lincoln: University of Nebraska Press, 1996), 206. Or, one might conclude from the present study, they may aim to be smarter when they need the cultures in which they dwell to be different.

19. George Steiner, "Some Meta-rabbis," in *Next Year in Jerusalem: Portraits of the Jew in the Twentieth Century* (New York: Viking Press, 1976), 64.

20. For a fascinating and sophisticated description of this resemblance, see Menachem Fisch, *Rational Rabbis: Science and Talmudic Culture* (Bloomington: Indiana University Press, 1997).

21. Linda Gradstein, "A Nobel Nation," *Jewish Exponent*, Oct. 12, 2011, www.jewishexponent.com/article/24542/A_Nobel_Nation.

22. Joseph Jacobs, "The Comparative Distribution of Jewish Ability," *Journal of the Anthropological Institute of Great Britain and Ireland* 15 (1886): 363.

23. Issac Deutscher, *The Non-Jewish Jew and Other Essays* (Oxford: Oxford University Press, 1968), 26. Deutscher attributed the success of these "non-Jewish Jews" to their marginality, in a way that reminds one of Veblen's theory. "They were *a priori* exceptional," he wrote,

> in that as Jews they dwelt on the borderlines of various civilizations, religions, and national cultures. They were born and brought up on the borderlines of various epochs. Their minds matured where the most diverse cultural influences crossed and fertilized each other. They lived on the margins or in the nooks and crannies of their respective nations. Each of them was in society and yet not in it, of it and yet not of it. It was this that enabled them to rise in thought above their societies, above their nations, above their times and generations, and to strike out mentally into wide new horizons and far into the future.

24. The quote appears in Hollinger, "Why Are Jews Preeminent in Science and Scholarship?," 158. See as well the remarkable essays gathered in David A. Hollinger, *Science, Jews, and Secular Culture: Studies in Mid-Twentieth-Century American Intellectual History* (Princeton, NJ: Princeton University Press, 1996), and in his *Cosmopolitanism and Solidarity: Studies in Ethnoracial, Religious, and Professional Affiliation in the United States* (Madison: University of Wisconsin Press, 2006).

25. Hollinger, *Cosmopolitanism and Solidarity*, 163.

26. Yuri Slezkine, *The Jewish Century* (Princeton, NJ: Princeton University Press, 2004), 207.

27. Slezkine, *Jewish Century*, 206.

28. For an influential discussion of the early history of these terms, see Salo W. Baron, "The Jewish Question in the Nineteenth Century," *Journal of Modern History* 10 (1938): 51–65.

29. This, for reasons that are easily seen. It was to "the Jewish Problem" that Hitler offered his "Final Solution," which was enough to squelch further public

discussion of a *Jewish* problem or question. But there were other reasons as well. In postwar Europe and America, the notion that there is a collective Jewish identity at all (much less a collective identity that stymies Jewish assimilation into the national societies in which they lived) came to seem less and less plausible. For overall, Jewish assimilation in Western democracies accelerated unabated in Western democracies. Just when it became unseemly to discuss the Jewish Problem, it became largely unnecessary.

30. Alfred Jay Nock, "The Jewish Problem in America," *Atlantic Monthly* 168 (June 1941): 699–706.

31. And, of course, not just Jews. The effort to fashion a *modus vivendi* with the majority culture was a hallmark of minority cultures in general, and especially minority and immigrant cultures.

32. Ernest Nagel, "Malicious Philosophies of Science," *Partisan Review* 10 (1943): 40–57.

33. Yaron Ezrahi, "Necessary Fictions: The Decline of Science in the Democratic Imagination," lecture given at Harvard University, Apr. 9, 2007. An excerpt is available at www.youtube.com/watch?v=oxz2rKwnDaU&feature=you tube_gdata_player (accessed Mar. 25, 2013).

Chapter 1. *"Holding High the Torch of Civilization"*

1. Indeed, New York rabbis had already assailed the Scopes trial in the weeks before it began. Speaking to the first graduating class of the Hebrew Union College School for Teachers, on June 2, 1925, Rabbi Samuel Schulman, chairman of the HUC Board of Governors, told the twenty-one graduates and their teachers: "There are dangerous signs in our country of a growing tendency to violate the fundamental principle of absolute separation of Church and State, which has been considered as established without question, and to prescribe by law certain forms of religious teaching and certain scientific theories. Judaism is absolutely opposed to such an attempt to force men's minds. . . . I cannot conceive it possible for any leader of Jewish thought to refer to scientists as 'dishonest scoundrels.' " "Twenty-One Get Diplomas in Jewish Training: First Class of Hebrew Union School for Teachers Is Graduated at Emanu-El," *New York Times*, June 3, 1925, 14.

2. "Rabbis Assail Law against Evolution: Defend Science in Rosh ha-Shanah Sermons, Denying It Destroys Religion. Condemn Modern Idols: Jews Urged to Check Peril of Fanaticism, Aid Progress and World Peace," *New York Times*, Sept. 20, 1925, E2. Unless otherwise noted, the rabbis cited in the subsequent text are all quoted in this detailed article.

3. Of all the rabbis, Schulman may have been the most sympathetic to fundamentalist Christians, saying that they "are guided by a correct instinct. They refuse all recognition of science. They are afraid. For the supreme dogma of science, its recognition of nothing else than visible nature, which can be investigated in its laboratories, will eventually destroy religion altogether." "Rabbis Assail Law against Evolution."

4. "Synagogues Packed as New Year Begins: Thousands of Jews Observe Rosh ha-Shanah Eve Here—Krass Tells of Israel's Mission. Schulman Stands by Bible, Wants Old Testament Upheld—Wise Praises Fosdick for Kindling Light in 'Sea of Unrest.'" *New York Times*, Sept. 19, 1925, 8.

5. "Rabbis Assail Law against Evolution."

6. Mac Davis, *From Moses to Einstein: They All Are Jews* (New York: Jordan Publishing, 1937), 7.

7. This genre remained popular for some time. One such book that I received for my own bar mitzvah was Ory Mazur, *Great Jews in Science* (New York: International Book Corp., 1968). This book's foreword begins:

> As citizens of the technological age in which we live, we are indebted to the pioneering work in the sciences of atomic energy, electronics, psychology and immunology—to mention but a few areas of contemporary scientific research which have improved our lives so greatly.
>
> Not a little of these vast extensions of scientific frontiers are the result of scientists of Jewish origin, whose roster, both in quantity and quality of achievement, bears eloquent testimony to the benefactions of the Jewish people to all humanity. The names of the better-known Jewish men of science have become household words of popular usage: Freudian psychology, Salk's polio vaccine, and Einstein's theory of relativity, conjure up, in the popular mind, the giants of the vast universe of science.

8. Statistics of this sort are crude and unreliable. It is difficult, perhaps impossible, to determine criteria that establish who is and is not Jewish (self-identification? Jewish heritage? traditional religious edicts?). Still, the basic trend that these crude numbers illustrate is undeniable: Jews—by any definition—have enjoyed success far greater than their raw numbers would have predicted. For thoughtful and persuasive discussions of this point, see Seymour Martin Lipset and Everett Carli Ladd, Jr., "Jewish Academics in the United States: Their Achievements, Culture, and Politics," *American Jewish Yearbook*, 1971, 89–128. Also see Paul Mendes-Flohr, *Divided Passions: Jewish Intellectuals and the Experience of Modernity* (Detroit, MI: Wayne State University Press, 1991), 23–53.

9. Richard E. Nisbett, *Intelligence and How to Get It: Why Schools and Cultures Count* (New York: Norton, 2010), 172.

10. Stephen Steinberg, *The Academic Melting Pot: Catholics and Jews in American Higher Education* (New York: McGraw-Hill, 1974).

11. Nisbett, *Intelligence and How to Get It*, 172.

12. John Simons, *Who's Who in American Jewry*, vol. 3 (New York: National News Association, 1938).

13. Shaul Katz, "On the Wings of the Brittle Rachis: Aaron Aaronsohn from the Rediscovery of Wild Wheat (Urwiezen) to His Vision for the Progress of Mankind," *Israel Journal of Plant Science* 49 (2001): S14. Among other things, Katz describes how Aaronsohn found support for his work through "the Jewish

philanthropic network" that was keen to support Jewish scientific research in Palestine.

14. "Address by Professor Einstein," *Science*, n.s., 77 (Mar. 17, 1933): 274–75.

15. Katz, "On the Wings of the Brittle Rachis," S14. Jacob Schiff also famously supported scientific endeavors among Jews in Palestine, such as the construction of the Technicum College in Haifa. Schiff was drawn to this project, as the *New York Times* reported, "because it seemed to offer a means for all Jews—German, Russian, and American, Orthodox, Reform, Zionist and anti-Zionist—to co-operate harmoniously in the cause of cultural elevation and progress in Palestine." He abandoned the project when some of its local advocates (Shmaryahu Levin in particular) insisted that the language of instruction be Hebrew, not German. See "Schiff Deplores Palestine Clash: Banker Doubts Whether American Jews Should Continue Support of the Technicum. Directors Soon to Meet, Will Confer in Berlin to Discuss Either Reorganization or Liquidation," *New York Times*, July 6, 1914. For the full breadth of Schiff's philanthropy, see the excellent book by Naomi W. Cohen, *Jacob H. Schiff: A Study in American Jewish Leadership* (Hanover, NH: Brandeis University Press, 1999).

16. Debi Unger and Irwin Unger, *The Guggenheims* (New York: Harper, 2005), 182–84.

17. This is, indeed, a very partial list. Writing in 1935, bacteriologist and chronicler of Jewish scientists Louis Gershenfeld gave a more (if still far from fully) complete list:

> In the United States, within recent years the list of Jews who have made large contributions for scientific endeavor is a long one. The Jew who pointed the way in days gone by and not content to rest but carried on until his recent death is that great crusader, a prince among givers, Nathan Straus, who represents in himself a great social service institution. But few big business men run their institutions with as much earnestness and efficiency as did the late Julius Rosenwald; his philanthropies and his great wealth enabled him to turn his lofty visions into realities. To these are to be added: The Guggenheim, Strauss, Warbug, Schiff, Loeb, Baruch, Lehman, Lewisohn, and Seligman families of New York City; the Bambergers and Fulds of Newark; the Falks, Kaufmans, Franks and Seders of Pittsburgh; Max Fleischmann of California; Nathan and Sol Lamport, Albert D. Lasker, Lucius Littauer, Mrs. Emanuel Mandel, F. N. Homberger, Max Adler and Max Epstein of Chicago; Isidore D. Morrison, and George Roos of California; A. Shuman and A. C. Ratshesky of Boston; J. Galland of Olympia, Washington; James Speyer, Morris Schnasi, Maurice L. Rothschild, Milton S. Florscheim, Albert B. Kupenheimer and Ernest Stein of Chicago; Odenheimer of New Orleans; Arther Joseph and the Fleischmann family of Cincinnati; Marck C. Steinberg and the Schoenberg family of St. Louis; Nathan Stone of Milwaukee; Leo J. Marks, Adolph Ochs, the Littauers, George Blumenthal, A. M. Hein-

sheimer, Mrs. Nathan Miller, Mrs. M. Hayman, Louis Marshall, James Speyer, F. L. Brown and Conrad Hubert, all of New York City; M. S. Fridenberg and the Mastbaum, Bloch, Feels, Lit, Gimbel and Fleisher families of Philadelpia; Alper of San Francisco; the Bachrachs of Atlantic City; Joseph Samuels of Providence, R.I.; Jacob Epstein of Baltimore; D. W. Simons of Detroit, and George Cohen of Texas.

Gershenfeld's list continues. See Louis Gershenfeld, *The Jew in Science* (Philadelphia: Jewish Publication Society of America, 1934), 218–19. For another, jauntier survey of the philanthropic efforts of American Jews before 1950, see Harry Golden and Martin Rywell, *Jews in American History: Their Contribution to the United States of America* (Charlotte, NC: H. L. Martin, 1950), 261–90.

18. This pattern had been set in place earlier. Already in 1891, Baron Maurice de Hirsch jumped into a celebrated debate about "the obligations of wealth," writing in the *North American Review*:

> Philanthropy in its proper sense has, no doubt, a higher purpose, and can find its best field for action in the creation of free libraries, green parks, beautiful churches, etc. This is Mr. Carnegie's idea, which he has practically demonstrated again and again. Certainly these are ideal objects, which tend to bring about universal happiness; and lucky are they who live in lands where the absolute necessaries of life are so well supplied. . . . In relieving human suffering I never ask whether the cry of necessity comes from a being who belongs to my own faith or not; but what is more natural than that I should find my highest purpose in bringing to the followers of Judaism, who have been oppressed for a thousand years, who are starving in misery, the possibility of a physical and moral regeneration than that I should try to free them, to build them up into capable citizens, and thus furnish humanity with much new and valuable material? Every page in the history of the Jews teaches us that in thinking this I am following no Utopian theory, and I am confident that such a result can be attained. ("My Views on Philanthropy," *North American Review* 416 [July 1891]: 1–2)

The commitment to projects of "universal value" that by their nature transform Jews into better citizens, more civil and productive, was seductive just as the large wave of Jewish immigrants began to arrive on America's shores, and remained so for generations after they arrived. For a valuable contemporary account of American Jewish Philanthropy, see Boris Bogen, *Jewish Philanthropy: An Exposition of Principles and Methods of Jewish Social Service in the United States* (New York: Macmillan, 1917). For a forceful description of Jewish philanthropy at the start of the twentieth century, filtered through a garnet lens of judaeophilia, see the following, by a one-time governor of Massachusetts and a member of Congress: Samuel W. McCall and Charles William Eliot, *Patriotism of the American Jew* (New York: Plymouth Press, 1924), 209–18.

19. "Jewish Institution Gets a WPA Mural: Three Panels by William Karp Accepted by Academy, Former Hebrew Orphan Asylum," *New York Times*, July 10, 1940.

20. The following example from Baltimore is typical: "Honors Slated for Dr. Macht: Hopkins Scientist Praised for 40 Years of Research," *The Sun (1837–1986)* (Baltimore), Dec. 18, 1949, 22.

21. Stephen M. Cohen provides a lovely description of Yiddish chemistry books published in America in "Chemical Literature in Yiddish: A Bridge between the Shtetl and the Secular World," *Aleph*, no. 7 (Jan. 1, 2007): 190–97.

22. Y. Blumshteyn, *Darvinizm Un Zayn Teorye*, (New York: Folksbildung, 1915), http://ia700404.us.archive.org/22/items/nybc204146/nybc204146.pdf.

23. From an advertisement at the end of Yankev Avrum Merison, *Fizyologye*, Arbeiter Ring Bibliotek (New York: Workmen's Circle, n.d.), 96, http://ia700409 .us.archive.org/8/items/nybc211418/nybc211418.pdf.

24. Solomon Herbert, *The First Principles of Evolution* (London: A. and C. Black, 1915), http://archive.org/details/cu31924003046509; published in Yiddish as *Evolutsye* (New York: Farlag di Heym, 1920), http://archive.org/details/nybc207134.

25. G. A. Guryev, *Darvinism un Atiezm* (Warsaw: Yatshkovski, 1931), http://ia700300.us.archive.org/9/items/nybc206689/nybc206689.pdf.

26. "Jewish Women Sponsoring Tour of Science Museum," *Chicago Daily Tribune*, Jan. 19, 1941, H6.

27. There is, of course, a story to tell about American Jews in the sciences before the start of the great migration of the 1880s. Its scale is surprisingly small, however. For one thing, "the Jewish population in the United States from 1840 to 1880 can properly be described as minuscule," as historian Rowena Olegario put it. Olegario, "'That Mysterious People': Jewish Merchants, Transparency, and Community in Mid-Nineteenth-Century America," *Business History Review* 73, no. 2 (July 1, 1999): 163. In fact, it is estimated that only about 15,000 Jews lived in the United States at the start of that period, and 250,000 by the end of the period. American Jewish Historical Society, *American Jewish Desk Reference* (New York, Random House: 1999), 35. The increase was accounted for, in the main, by immigrants from German-speaking lands, who overwhelmingly sought their livelihoods in commerce. Hasia R. Diner, *A Time for Gathering: The Second Migration, 1820–1880* (Baltimore: Johns Hopkins University Press, 1992), 11. Add to this the fact that the United States was itself a scientific backwater for a good part of the nineteenth century, and it is perhaps no surprise that in numbers, impact, and reputation, American Jews were a diminutive force in science in this period. (For an excellent, brief survey of the state of American science in the nineteenth century, see Karen Hunger Parshall, "America's First School of Mathematical Research: James Joseph Sylvester at the Johns Hopkins University 1876–1883," *Archive for History of Exact Sciences* 38, no. 2 [Jan. 1, 1988]: 155–61.)

Despite all this, there were Jews who made names in the sciences during this period. Writing from the vantage of 1934, Louis Gershenfeld tirelessly catalogued Jews who had been active in American sciences and found several dozen who had been born before 1875. Among these were Alfred Abraham Michelson, a German émigré physicist who in time was awarded a Nobel Prize, and Julius Oscar Stieglitz, the son of a German emigrant, who completed his own education as a chemist in Germany before assuming a post at the University of Chicago. Gershenfeld, "American Jewish Scientists," in *The Jew in Science*, 167–204. One especially moving and illustrative example is that of British émigré mathematician James Joseph Sylvester, who fled the United Kingdom to escape a manslaughter charge, and eventually founded the department of mathematics at the newly established Johns Hopkins University. Sylvester gave full-throated voice to his frustration with the bigotries that he encountered in the New World. On the disappointments Sylvester suffered in America, see Karen Hunger Parshall, *James Joseph Sylvester: Jewish Mathematician in a Victorian World* (Baltimore: Johns Hopkins University Press, 2006), 69–80. See, too, Nathan Reingold, ed., *The Papers of Joseph Henry*, vol. 5 (Washington, DC: Smithsonian Institution Press, 1985), xv, 92, 359, 362, 366, 369–70; Marc Rothenberg, ed., *The Papers of Joseph Henry*, vol. 6 (Washington, DC: Smithsonian Institution Press, 1992), 380.

28. John Higham, *Send These to Me: Jews and Other Immigrants in Urban America* (New York: Atheneum, 1975); John Higham, "American Anti-Semitism Historically Reconsidered," in *Jews in the Mind of America* (New York: Basic Books, 1966), 237–58. Also see Milton Goldin, *Why They Give: American Jews and Their Philanthropies* (New York: Macmillan, 1976), 50–51.

29. John Higham, "Anti-Semitism in the Gilded Age: A Reinterpretation," *Mississippi Valley Historical Review* 43 (1957): 563–64.

30. Higham, "Anti-Semitism in the Gilded Age," 573.

31. Bryan Cheyette, *Constructions of "the Jew" in English Literature and Society: Racial Representations, 1875–1945* (Cambridge: Cambridge University Press, 1995), 6.

32. *New York Sun*, Mar. 24, 1895, cited in Higham, "Anti-Semitism in the Gilded Age," 574.

33. William Dean Howells, "An East-Side Ramble," in *Impressions and Experiences* (New York: Harper, 1896), 127–49.

34. Jacob Riis, "The Jews of New York," *Review of Reviews* 13 (1896): 58–62, http://tenant.net/Community/LES/jacob4.html.

35. Howells, "East-Side Ramble," 148–49.

36. Riis, "Jews of New York," 59.

37. Ronald Sanders, *The Downtown Jews: Portraits of an Immigrant Generation* (New York: Harper & Row, 1969), chaps. 3–4.

38. Shalom Aleichem, "We Strike," in Henry Goodman, *The New Country: Stories from the Yiddish about Life in America* (New York: Yidisher Kultur Farband, 1961), 85, cited in Paul Buhle, *From the Lower East Side to Hollywood: Jews in American Popular Culture* (London: Verso, 2004), 28.

39. For an exhaustive and enchanting contemporary catalogue of efforts to establish Jewish farms, homesteads, and cooperatives, see Leonard George Robinson, *The Agricultural Activities of the Jews in America* (New York: American Jewish Committee, 1912).

40. "East Side in Tears as London Is Buried: Solid Mass of Humanity Packs Streets to Pay Tribute to Socialist and Labor Leader," *New York Times*, June 10, 1926, 25.

41. Abraham Cahan, "The Russian Jew in America," *Atlantic Monthly* 82 (1898): 263.

42. The difference in the cultural patina of well-established Jews of German origin and greenhorn eastern Europeans was considered with concern not only by Jews. Burton Hendrick, a writer with two Pulitzer Prizes and an enthusiastic following, published a book-length warning about the hordes of "Polish" Jews (by which he meant *Ostjuden*) who had darkened the docks, streets, and slums of New York:

> The process of "Americanization" is going to be slow and more difficult with this class of immigrants than with any other, except perhaps the Southern Italians. The million and a half Jews—probably more—that crowded into the New York tenements are by no means digested. This does not mean they never will be. The completeness with which the Sephardic and German Jews have been absorbed into the national life, the high standard of patriotic citizenship they have evinced, and the services they have rendered in the education, science, letters and other activities, show that there is nothing in the Jewish nature that necessarily dooms him to forever be alien. But the task with the Polish Jew is going to be a longer and more difficult one. That there are hundreds of thousands in New York's Polish Jew population who do not align themselves with these radical movements, and to whom American history and American institutions mean much, is clear. The unpleasant fact, however, is that there is an uncomfortable remainder who do.

Hendrick's conclusion was this: "There is no more hopeful manifestation in American life to-day than the fact that the Nation, after fifty years of fumbling and discussing, has at last reached the point of wisdom in the restriction of immigration." Burton Jesse Hendrick, *The Jews in America* (New York: Doubleday/Page, 1923), 170–71.

43. The book that best captures this phenomenon in a somewhat later recension is Norman Podhoretz's searing *Making It* (New York: Random House, 1967).

44. Allen Bodner, *When Boxing Was a Jewish Sport* (Westport, CT: Praeger, 1997), 9–12. A marvelous meditation on the relationship between Jewish identity and boxing can be found in Douglas Century's *Barney Ross* (New York: Schocken, 2009).

45. Gerald R. Gems, "Sport and the Forging of a Jewish-American Culture," *American Jewish History* 83, no. 1 (Mar. 1995): 23.

46. The remarkable story of Jewish domination of American basketball in the first half of the twentieth century is recounted in Arieh Schlar's engaging dissertation, "'A Sport at Which Jews Excel': Jewish Basketball in American Society, 1900–1951" (Stony Brook University, Stony Brook, NY, 2008).

47. For the place of sport in the formation of American Jewish identity, see George Eisen, "Jewish History and the Ideology of Modern Sport: Approaches and Interpretations," *Journal of Sport History* 25, no. 3 (1998): 482–531; Jeffrey S. Gurock, *Judaism's Encounter with American Sports* (Bloomington: Indiana University Press, 2005); and Peter Levine, *Ellis Island to Ebbets Field: Sport and the American-Jewish Experience* (New York: Oxford University Press, 1992).

48. The comparison of athletes and scientists was not lost on American Jews at the time. An article in 1925 in the *Jewish Bulletin* found that between Benny Leonard and Albert Einstein, the boxer was the greater Jew; both were known by millions, the article explained, but only the boxer was *understood* by these multitudes. See Gems, "Sport and the Forging of a Jewish-American Culture," 23; and Jack Kugelmass, "Why Sports?," in *Jews, Sports, and the Rites of Citizenship* (Urbana: University of Illinois Press, 2007), 24–25.

49. Heywood Broun and George Britt, *Christians Only: A Study in Prejudice* (New York: Vanguard Press, 1931), 231.

50. Daniel J. Kevles, *The Physicists: The History of a Scientific Community in Modern America* (Cambridge, MA: Harvard University Press, 1987), 74.

51. Nathan Reingold and Ida Reingold, *Science in America: A Documentary History, 1900–1939* (Chicago: University of Chicago Press, 1981), 14–16.

52. In 1911, *Science* reported with enthusiasm the founding of the new Kaiser-Wilhelm Institut für Physikalische Chemie und Elektrochemie, to be headed by Fritz Haber, scion of a Jewish family of some renown. The notice emphasized that "the director has an absolutely free hand in the choice of his work, his fellow workers and his assistants. . . . There are no restrictions whatever as to the nationality of the men admitted by the director." This lack of attention to background was an integral element of the ideology of the new Kaiser Wilhelm Institutes and the culture of science they were meant to spawn. See William D. Harkins, "The 'Kaiser-Wilhelm Institut für Physikalische Chemie und Elektrochemie,'" in *Science*, n.s., 34 (Nov. 3, 1911): 595–97.

For an excellent account of Taylor and Taylorism in their day, see Robert Kanigel, *The One Best Way: Frederick Winslow Taylor and the Enigma of Efficiency* (Cambridge, MA: MIT Press, 2005).

53. See David Hull, "Openness and Secrecy in Science: Their Origins and Limitations," *Science, Technology, & Human Values* 10, no. 2 (Spring 1985): 4–13.

54. "Resents Rejection of Loeb: Professor Cattell Charges That Distinguished Scientist Was Blackballed by Century Club Because He Is a Jew," *Boston Evening Transcript*, May 14, 1913, 7, available online at http://news.google.com/newspape

rs?nid=2249&dat=19130514&id=Vbk-AAAAIBAJ&sjid=9lkMAAAAIBAJ&pg
=1389,2475964 (accessed Aug. 2, 2012). An anonymous member of the admissions committee insisted that Loeb was rejected from the club not because he was Jewish but because he had a "strong predilection for socialism, . . . is otherwise erratic in his views, emphatic in his expression of them and not always tolerant of those who oppose them." Quoted in "Rejection of Loeb Stirs Century Club: Friends Say Committee Blackballed Him Because He Is a Jew—Prof. Cattell Is Angry—Says the Scientist Is One of the Great Men of the World," *Fulton Sun*, May 14, 1913, 6.

The convictions that led Catell to challenge the Century Club remained close to his heart. In 1927, he published a vigorous essay entitled "Science, the Declaration, Democracy" in which he stated: "It is science that has led the way to true freedom of thought. . . . Those with the power, whether in state or church, have ever been ready to impose their creeds and their control upon others. . . . The methods of science, slowly gaining in force and volume through the centuries, will in the end bring truth and reason into all our beliefs and actions." J. McKeen Cattell, "Science, the Declaration, Democracy," *Scientific Monthly* 24, no. 3 (Mar. 1, 1927): 201.

55. "Loeb on Racism, Zionism and Ethnocentrism," in Charles Rasmussen and Rick Tilman, *Jacques Loeb: His Science and Social Activism and Their Philosophical Foundations* (Philadelphia: American Philosophical Society, 1998), 137.

56. Philip J. Pauly, *Controlling Life: Jacques Loeb and the Engineering Ideal in Biology* (New York: Oxford University Press, 1987), 130.

57. "Zionist Leaders Greeted by Vast Throng at the Arrival in New York," *Jewish Independent*, Apr. 8, 1921, 1, reprinted in the enchanting compendium, József Illy, ed., *Albert Meets America: How Journalists Treated Genius during Einstein's 1921 Travels* (Baltimore: Johns Hopkins University Press, 2006), 25. A Jewish paper called *The New Palestine* reported that "over 10,000 people were at the dock to see [Einstein and Weizmann] land, and so great was the joy of welcome that they followed the automobile that took the group to the Hotel Commodore, and refused to disperse until far into the night." Reprinted in Illy, *Albert Meets America*, 71.

58. Quoted in Illy, *Albert Meets America*, xv.

59. "Huge Throng Welcomes Visiting Zionist Leaders," *Jewish Independent*, Apr. 15, 1921, 1–5, reprinted in Illy, *Albert Meets America*, 85.

60. The essay that spurred Cohen to write was "Carpentier: A Symbol," *New Republic*, July 20, 1921, 206. Cohen's reply appeared as "Mr. Einstein: A Correction," *New Republic*, Aug. 24, 1921, 356. Both pieces are reprinted in Illy, *Albert Meets America*, 331–32.

61. Morris Raphael Cohen, "The Jew in Science," in *Reflections of a Wondering Jew* (New York: Free Press, 1950), 99–100. The Judaeans were a remarkable group of Jewish intellectuals and power brokers who met regularly for more than a generation. For a sense of the sorts of issues that occupied them, see Philip

Cowen, *The Judaeans, 1897–1899* (New York: Society Press, 1899), and Henry Leipziger, *Judaean Addresses*, vol. 2, *1900–1917* (New York: Bloch, 1917).

62. Morris Raphael Cohen, "What I Believe as an American Jew," in *Reflections of a Wondering Jew*, 3.

63. Hyman Levy, *The Universe of Science* (New York: Century, 1933), 189.

64. Robert King Merton, "Science and the Social Order," *Philosophy of Science* 5, no. 3 (July 1938): 326–27. It is interesting to note the similarities and differences between Merton's formulation and that of Watson Davis, the director of the Science Service (a nonprofit organization started in 1921 to educate a wide public about science), who saw the six elements of a scientific attitude as accuracy, intellectual honesty, open-mindedness, suspended judgment, looking for true cause-and-effect relationships, and criticism, including self-criticism. See Peter J. Kuznick, *Beyond the Laboratory: Scientists as Political Activists in 1930's America* (Chicago: University of Chicago Press, 1987), 46. Merton's account adds the social virtues of disinterestedness and impersonality. For a superlative account of how Merton's philosophy of science bore upon his political views (and vice versa), see David A. Hollinger, "The Defense of Democracy and Robert K. Merton's Formulation of the Scientific Ethos," *Science, Jews, and Secular Culture: Studies in Mid-Twentieth-Century American Intellectual History* (Princeton, NJ: Princeton University Press, 1996), 80–96.

65. Quoted in "Nazi's Conception of Science Scored: 1,284 American Scientists Sign Manifesto Rallying Savants to Defend Democracy," *New York Times*, Dec. 11, 1938, 50. The quotation within the text is taken from a resolution passed by the American Association for the Advancement of Science in December 1938, condemning Nazi discrimination against non-Aryan scientists.

66. As told by Franz Boas, "Race Prejudice from the Scientist's Angle," *Forum and Century*, Aug. 1937, 94.

67. Indeed, Boas' manifesto gained traction in institutions that were themselves still well-known for placing their own restrictions on the scientific education and career advancement of Jews. See, for instance, "Resolution Urges Scientists to War On Fascist Forces," *Harvard Crimson*, www.thecrimson.com/article/1938/12/13/resolution-urges-scientists-to-war-on, accessed Aug. 26, 2012.

68. Andrew Jewett, *Science, Democracy, and the American University* (Cambridge: Cambridge University Press, 2012), vii, 2, 10.

69. Jewett, *Science, Democracy and the American University*, 10.

70. Jewett himself finds Jews prominent among the opponents of "scientific democracy" as well, and none more than Merton. "Led by Columbia's Robert K. Merton," Jewett writes, "postwar sociologists of science focused on explicating the internal dynamics of a freestanding 'scientific community,' which they portrayed as set off entirely from the surrounding society. The field-defining sociological accounts of the World War II and early Cold War years stressed science's institutional autonomy, not its cultural matrix." Jewett, *Science, Democracy and the*

American University, 243. According to Jewett, in Merton's view "only science's near-total isolation from other social institutions—above all, from the political system—ensured its success" (247). Jewett is right, of course, that Merton insisted that the autonomy of science had to be jealously safeguarded from political intervention of the crass sort practiced by the Nazis. But he believed equally that the ethos that governed the autonomous realm of science—intellectual honesty, integrity, organized skepticism, disinterestedness, impersonality—were values that could and should gain greater purchase in public life in general, and that science was a vehicle for spreading this ethos in the public square.

71. Golden and Rywell, *Jews in American History*, 297.

72. The tradition of seeing in public schools institutions whose fundamental purpose is advancing an ideal of separation of church and state, and a commitment to meritocracy blind to the creed and ancestry, must be part of any explanation for the vigorous and very nearly unanimous rejection over the past decades of teaching "creation science" or "intelligent design" in public schools. For a detailed explanation of why this is so, see Noah Efron, "American Jews and Intelligent Design," Reilly Center Reports 1, http://reilly.nd.edu/assets/65756/rcrefron.pdf, accessed Mar. 20, 2013.

73. Louis I. Newman, *Vote "No" On Amendment No. 17: Why Bible Reading Should Not Be Required in the Public Schools of California* (San Francisco: General Committee, District Lodge No. 4, I.O.B.B., 1926).

74. Quoted in Louis I. Newman, *The Sectarian Invasion of Our Public Schools* (San Francisco, 1925), 43–44.

75. Newman, *Sectarian Invasion*, 47.

76. Newman, *Sectarian Invasion*, 46.

77. As was true in the later case of broad Jewish support for the civil rights movement. About this, see my "All We Did for Them," *Boston Book Review*, Nov. 1997, 18–19.

78. My discussion of the judicial development of the establishment clause is much indebted to Gregg Ivers' extraordinary book *To Build a Wall: American Jews and the Separation of Church and State, Constitutionalism and Democracy* (Charlottesville: University Press of Virginia, 1995).

79. This story is wonderfully told, with great care and detail, by Ivers, *To Build a Wall*.

80. As I have written elsewhere:

No American Jewish organization advocates teaching intelligent design in public schools. No Jewish religious body endorses it. Official representatives of all the major streams of Judaism—Reform, Conservative, Orthodox & Reconstructionist—have denounced it; only Ultra-Orthodox Jews, who in any case vigilantly oppose sending Jewish children to public school, have expressed sympathy for teaching intelligent design in the schools they refuse to attend, and even then in very small numbers. Jewish organizations like the American Jewish Committee supply attorneys to prosecute school boards introducing intelligent design into their curriculum, and split the tab for the ensuing trials. Among their non-denominational partners—the ACLU, People for the American Way, and others—Jews are represented in the rank and file in numbers that vastly exceed what one might expect of a small minority. In a famously fractious community—an old joke has two Jews stranded alone on a dessert island; the sea captain who finally rescues them is perplexed to find they've built three synagogues—almost all American Jews agree that intelligent design has no place in the public school science curriculum. And polls consistently show that Jews hold this view far more commonly than members of any other religious or ethnic group in America. (Efron, "American Jews and Intelligent Design," 1)

81. Benjamin Charles Gruenberg, *Science and the Public Mind* (New York: McGraw-Hill, 1935), 28–40.

82. Benjamin C. Gruenberg, *The Story of Evolution* (Garden City, NY: Garden City Publishing, 1929), 460.

83. Benjamin C. Gruenberg, "Teaching Biology after the Wars," *American Biology Teacher* 9, no. 4 (Jan. 1947): 104.

84. Sidonie Matsner Gruenberg and Benjamin C. Gruenberg, "Crosscurrents in the Rearing of Youth," *Annals of the American Academy of Political and Social Science* 236 (Nov. 1944): 69, 73.

85. Quoted in Marianne R. Sanua, "Jewish College Fraternities in the United States, 1895–1968: An Overview," *Journal of American Ethnic History* 19, no. 2 (Jan. 1, 2000): 11–12.

86. For a telling comparison, brilliantly told, between two Jewish physicists of mythic reputations, see S. S. Schweber, "Intersections: Oppenheimer and Einstein," in *Reappraising Oppenheimer: Centennial Studies and Reflections*, ed. Cathryn Carson and David A. Hollinger (Berkeley: Office for History of Science and Technology, University of California, 2005), 343–60; and S. S. Schweber, *Einstein and Oppenheimer: The Meaning of Genius* (Cambridge, MA: Harvard University Press, 2008).

87. Their common interest in physics bridged the gap between the two men, apparently, but without closing it. Rabi told fellow physicist Jeremy Bernstein this of Oppenheimer:

> I found him excellent. We got along very well. We were friends until his last day. I enjoyed things about him that some people disliked. It's true you carried on a charade with him. He lived a charade, and you went along with it. . . . He reminded me of a boyhood friend about whom someone said that he couldn't make up his mind whether to be president of the B'nai B'rith or the Knights of Columbus. Perhaps he wanted to be both simultaneously. Oppenheimer wanted every experience. In that sense he never focused. My own feeling is that if he had studied the Talmud

and Hebrew, rather than Sanskrit, he would have been a much greater physicist. (Jeremy Bernstein, "Oppenheimer's Beginnings," *New England Review* 25, nos. 1/2 [Jan. 1, 2004]: 38)

88. I. I. Rabi, *Science: The Center of Culture* (Omaha: World Publishing, 1970), 45.

89. J. Robert Oppenheimer, *Science and the Common Understanding* (New York: Simon and Schuster, 1954), 7. A riveting recording of a portion of the lecture series is available at http://downloads.bbc.co.uk/podcasts/radio4/rla48/rla48_19531220-0900b.mp3.

90. Robert J. Oppenheimer, *Atom and Void: Essays on Science and Community* (Princeton, NJ: Princeton University Press, 1989), 13–14.

91. Oppenheimer's views on the role of science in society were complex and changed over time. Early in his career, he saw in science (at least at times) a model for modern, liberal, democratic politics. Over the years, and especially after he was stripped of his security clearance, he emphasized the gap between science as an activity best carried out by a cloistered elite and what he saw as increasingly messy mass culture and mass politics. The sciences have become "encysted and broken off from the great stream . . . of our common life," Oppenheimer said in a lecture at St. Andrew's University. "The situation today, in fact, seems to involve an alienation between the world of science and the world of public discourse." J. Robert Oppenheimer, *Some Reflections on Science and Culture* (Chapel Hill: University of North Carolina, 1960), 8. For an excellent description of the subtleties of Oppenheimer's views of the relationships between science and politics, and of how they changed with the years, see Charles Thorpe, "The Scientist in Mass Society: J. Robert Oppenheimer and the Postwar Liberal Imagination," in Carson and Hollinger, *Reappraising Oppenheimer*, 293–314.

92. Oppenheimer remains a figure of undiminished fascination to historians, and for good reason. A man of rare gifts, he seems at once to embody and to stand apart from the history of American physics in arguably its most heroic epoch. "In all my life I have never known a personality more complex than Robert Oppenheimer," wrote physicist and historian Abraham Pais in his biography, *J. Robert Oppenheimer: A Life* (New York: Oxford University Press, 2006), 139. Perhaps as a result, almost no aspect of Oppenheimer's life, work, and thought is innocent of controversy. "What is so special about Oppenheimer," David Hollinger felt, "is the concentration in a single individual's life of so many of his era's most important pressures and novelties." David A. Hollinger, "Afterword," in Carson and Hollinger, *Reappraising Oppenheimer*, 386. For excellent recent scholarship, also see Jeremy Bernstein, *Oppenheimer: Portrait of an Enigma* (Chicago: Ivan R. Dee, 2004); Kai Bird, *American Prometheus: Triumph and Tragedy of Robert Oppenheimer* (New York: Alfred A. Knopf, 2006); David C. Cassidy, *J. Robert Oppenheimer and the American Century* (New York: Pi Press, 2005); Ray Monk, *Robert Oppenheimer: A Life inside the Center* (New York: Random House, 2013; and

Jennet Conant, *109 East Palace: Robert Oppenheimer and the Secret City of Los Alamos* (New York: Simon & Schuster, 2005).

Chapter 2. *"Second Only to Communism"*

1. Michael Beizer, "OZE" in *YIVO Encyclopedia of Jews in Eastern Europe* (2010), www.yivoencyclopedia.org/article.aspx/OZE (accessed Mar. 31, 2013). Also see Nadav Davidovitch and Rakefet Zalashik, "'Air, Sun, Water': Ideology and Activities of OZE during the Interwar Period," *Dynamis* 28 (2008): 127–49; Jacob Joshua Golub, "OSE: Pioneer of Jewish Health," *Jewish Social Service Quarterly* 14, no. 4 (June 1938): 3–16; Lazar Gourvitch, *Twenty-Five Years, OSE, 1912–1937* (Paris: OSE, 1937); and Jacob Lestschinsky, *OSE: 40 Years of Activities and Achievements* (New York: American Committee of OSE, 1952).

2. Robert Weinberg, "Biology and the Jewish Question after the Revolution: One Soviet Approach to the Productivization of Jewish Labor," *Jewish History* 21, nos. 3/4 (2007): 413–28.

3. Quoted in Deborah Hope Yalen, "Red Kasrilevke: Ethnographies of Economic Transformation in the Soviet Shtetl, 1917–1939" (PhD diss., University of California, Berkeley, 2007), 128.

4. Theodore H. Friedgut and Bella Kotik-Friedgut, "A Man of His Country and His Time: Jewish Influences on Lev Semionovich Vygotsky's World View," *History of Psychology* 11, no. 1 (2008): 15–39.

5. R. Luria, *Cognitive Development: Its Cultural and Social Foundations* (Cambridge, MA: Harvard University Press, 1976).

6. A. R. Luria, "Vygotsky and the Problem of Functional Localization" in *The Selected Writings of A. R. Luria*, ed. M. Cole (New York: Sharpe, 1978), 275.

7. Lev Shternberg, for instance, who served as the senior curator of the Museum of Anthropology and Ethnography in Saint Petersburg/Leningrad from 1901 to 1927, as anthropology editor for the canonical Brockhaus and Efron Russian encyclopedia, and after the revolution as the founder (along with another Jewish anthropologist named Vladimir Bogoraz) of the Leningrad school of anthropology. About this remarkable person's life work, see Sergei Kan, *Lev Shternberg: Anthropologist, Russian Socialist, Jewish Activist* (Lincoln: University of Nebraska Press, 2009).

8. Francine Hirsch, *Empire of Nations: Ethnographic Knowledge and the Making of the Soviet Union, Culture and Society after Socialism* (Ithaca, NY: Cornell University Press, 2005), 8–9.

9. Wendell Philips, "Speech at the Melodeon, Wednesday Evening, January 28," in *Speeches before the Massachusetts Anti-Slavery Society* (Boston: Robert F. Wallcut, 1852), 4.

10. Quoted in Yuri Slezkine, *The Jewish Century* (Princeton, NJ: Princeton University Press, 2004), 124.

11. Quoted in Benjamin Nathans, *Beyond the Pale: The Jewish Encounter with Late Imperial Russia* (Berkeley: University of California Press, 2004), 205.

12. Nathans, *Beyond the Pale*, 207.

13. Quoted in Nathans, *Beyond the Pale*, 215–16.

14. Salo Wittmayer Baron, *The Russian Jew under Tsars and Soviets* (New York: Schocken Books, 1987), 137; Nathans, *Beyond the Pale*, 102–3.

15. A government memo explaining the quotas stated: "Individuals of Polish and Jewish descent, having a particular capacity for mathematics, are more successful than others on the entrance exams. . . . Every year the percentage of non-Russians among mining engineers increases and . . . a large portion of the management positions in factories and mines will pass into the hands of individuals of Polish and Jewish origin." Nathans, *Beyond the Pale*, 267–8.

16. Nathans, *Beyond the Pale*, 263.

17. Chaim Weizmann, *Trial and Error: The Autobiography of Chaim Weizmann* (Philadelphia: Jewish Publication Society of America, 1949), 19.

18. Nathans, *Beyond the Pale*, 272. The newly installed quotas were not always followed in full. Eighteen percent of students at Kiev University in 1899 were Jews, even though the official quota limited them to 10%. By law, Moscow University was allowed 336 Jewish students in 1916, but in fact, 684 were enrolled. The quota for the university in Saint Petersburg limited the number of Jewish students to 3% of the total; in 1915, 11% of the students there were Jews. Alexander Vucinich, *Science in Russian Culture, 1861–1917* (Stanford, CA: Stanford University Press, 1971), 376. But while these numbers of students exceeded the legal limits, they were much smaller than the number of Jews who would have attended university had there been no quotas at all.

19. Omeljan Pritsak, "The Pogroms of 1881," *Harvard Ukrainian Studies* 11, nos. 1/2 (June 1, 1987): 8–43.

20. "Jewish Massacre Denounced," *New York Times*, April 28, 1903, 6.

21. For a good discussion of the immediate context of the waves of anti-Semitic attacks in late Imperial Russia, see Shlomo Lambroza, "The Pogrom Movement in Tsarist Russia, 1903–6" (PhD diss., Rutgers University, 1981). Also see John Doyle Klier and Shlomo Lambroza, eds., *Pogroms: Anti-Jewish Violence in Modern Russian History* (Cambridge: Cambridge University Press, 2004).

22. Nathans, *Beyond the Pale*, 377. Indeed, to these "two Russian Jewries" a third may be added: the great number of Jews who traveled to western Europe to pursue an education denied to them by Russian university quotas. In 1912, V. I. Vernadskii estimated that there were "at least 7,000 to 8,000" such Jews and that they were the major share of all Russians studying abroad. As I describe later in this chapter, many of these young scholars, most of them scientists, would find places of influence in Soviet science when they eventually returned to their homeland. Vucinich, *Science in Russian Culture*, 377.

23. Steven Cassedy, "Russian-Jewish Intellectuals Confront the Pogroms of 1881: The Example of 'Razsvet,'" *Jewish Quarterly Review* 84, nos. 2/3 (Oct. 1, 1993): 129–52.

24. Nathans, *Beyond the Pale*, 379. For a more detailed discussion of Levin

and the other pioneering Jewish deputies in the Duma, see Sidney S. Harcave, "The Jewish Question in the First Russian Duma," *Jewish Social Studies* 6, no. 2 (Apr. 1, 1944): 155–76.

25. David Shavit, "The Emergence of Jewish Public Libraries in Tsarist Russia," *Journal of Library History* 20, no. 3 (July 1, 1985): 239–52.

26. Abraham Cahan, "Jewish Massacres and the Revolutionary Movement in Russia," *North American Review* 177, no. 560 (July 1, 1903): 49–62.

27. From the poem "Rusland," quoted in Nathans, *Beyond the Pale*, 367.

28. Nathans, *Beyond the Pale*, 378.

29. Alexei Kojevnikov, "The Great War, the Russian Civil War, and the Invention of Big Science," *Science in Context* 15, no. 2 (June 2002): 249.

30. Vucinich, *Science in Russian Culture*, 103.

31. Elizabeth A. Hachten, "In Service to Science and Society: Scientists and the Public in Late-Nineteenth-Century Russia," *Osiris* 17 (2002): 209.

32. Samuel D. Kassow, *Students, Professors, and the State in Tsarist Russia* (Berkeley: University of California Press, 1989), 204.

33. Kojevnikov, "The Great War," 241.

34. Petr Alekseevich Kropotkin, *An Appeal to the Young* (London: "Justice" Printery, 1889), 6.

35. Kojevnikov, "The Great War," 242.

36. Kassow, *Students, Professors, and the State*, 220.

37. Kassow, *Students, Professors, and the State*, 288.

38. Kojevnikov, "The Great War," 248.

39. Kojevnikov, "The Great War," 239–40.

40. Quoted in Kojevnikov, "The Great War," 251–52.

41. Vucinich, *Science in Russian Culture*, 380.

42. Quoted in Alexei B. Kojevnikov, *Stalin's Great Science: The Times and Adventures of Soviet Physicists* (Hackensack, NJ: World Scientific Publishing, 2004), 7–8.

43. Quoted in Kojevnikov, "The Great War," 253.

44. Fred P. Haggard, "The New Spirit in Russia," *Journal of Race Development* 8, no. 3 (Jan. 1, 1918): 288.

45. Vladimir Lenin, "Anti-Jewish Pogroms," *Lenin's Speeches on Gramophone Records*, www.marxists.org/romana/audio/speeches/antisem.htm (accessed Sept. 1, 2012). This website provides the original speech as an audio file as well as a Russian-language transcription and an English translation.

46. Yuri Slezkine, *The Jewish Century*, 222–23.

47. Arthur Ruppin, *The Jewish Fate and Future* (London: Macmillan, 1940), 272.

48. Louis Harap, "More Comments on Howard Fast," *Masses & Mainstream*, Apr. 1957, 55.

49. Shmuel Adler, "Job Training and Retraining Programs in Israel," in *Fourth Metropolis International Conference, Workshop on Barriers to Employment Faced by Immigrants* (Washington, DC, 1999).

50. Raphael Patai, *The Jewish Mind* (New York: Scribner, 1977), 341.

51. Edward Teller, "Some Personal Memories of George Gamow," George Gamow Symposium, 1997, Astronomical Society of the Pacific Conference Series, vol. 129, ed. E. Harper, W. C. Parke, and D. Anderson, 125; available online at http://adsabs.harvard.edu/abs/1997ASPC..129..123T (accessed Apr. 4, 2013).

52. Erwin N. Hiebert, "Abram Fedorovič Ioffe: Vater der Sowjetischen Physik," *Isis* 83, no. 1 (Mar. 1992): 159.

53. Vucinich, *Science in Russian Culture*, 377. See also Horst Kant, *Abram Fedorovic Ioffe: Vater der Sowjetischen Physik (Biographien Hervorragender Naturwissenschaftler, Techniker und Mediziner)*, 1. Aufl (Weisbaden: B. G. Teubner, 1989).

54. Krzysztof Szymborski, "The Physics of Imperfect Crystals: A Social History," *Historical Studies in the Physical Sciences* 14, no. 2 (Jan. 1, 1984): 317–55.

55. Vucinich, *Science in Russian Culture*, 377. Also see Paul Josephson, *Physics and Politics in Revolutionary Russia* (Berkeley: University of California Press, 1991), 30–31.

56. Lev Theremin, "Erinnerungen an A. F. Joffe," translated from the Russian by Felix Eder, www.ima.or.at/theremin/?page_id=56 (accessed Oct. 2, 2012).

57. Josephson, *Physics and Politics*, 31.

58. Szymborski, "The Physics of Imperfect Crystals."

59. Michael Parrish, *Sacrifice of the Generals: Soviet Senior Officer Losses, 1939–1953* (Lanham, MD: Scarecrow Press, 2004), 264.

60. Szymborski, "The Physics of Imperfect Crystals," 325.

61. Josephson, *Physics and Politics in Revolutionary Russia*, 97.

62. On this exile of intellectuals and its causes and effects, see the remarkable study by Lesley Chamberlain, *Lenin's Private War: The Voyage of the Philosophy Steamer and the Exile of the Intelligentsia* (New York: St. Martin's Press, 2007).

63. V. I. Lenin, "On the Significance of Military Materialism," in *The Lenin Anthology*, ed. Robert Tucker (New York: Norton, 1975), 651–53. On the perceived affinities between the ideology of the Soviet revolution and that of modern science, see as well Loren R. Graham, "Bukharin and the Planning of Science," *Russian Review* 23, no. 2 (Apr. 1, 1964): 135–48.

64. Josephson, *Physics and Politics*, app. B, tables 4–5.

65. James Gerald Crowther, *Science in Soviet Russia* (London: Williams & Norgate, 1930), 42.

66. Crowther, *Science in Soviet Russia*, 42.

67. For a detailed contemporary description of the Kharkov Institute, its faculty, and their research, see Crowther, *Science in Soviet Russia*, 77–105. A retrospective look at the institute's activities is provided in V. E. Ivanov and V. F. Zelenskii, "Fiftieth Anniversary of the Kharkov Physicotechnical Institute of the Academy of Sciences of the Ukrainian SSR," *Atomic Energy* 45, no. 4 (1978): 1008–17.

68. Isaak M. Khalatnikov, *From the Atomic Bomb to the Landau Institute:*

Autobiography; Top Non-Secret (Heidelberg: Springer, 2012), 9; Crowther, *Science in Soviet Russia*, 126–41.

69. On the fascinating and ultimately tragic figure of Boris Hessen see Gideon Freudenthal and Peter McLaughlin, eds., *The Social and Economic Roots of the Scientific Revolution: Texts by Boris Hessen and Henryk Grossmann* (Heidelberg: Springer, 2009), especially Freudenthal and McLaughlin, "Boris Hessen: In Lieu of a Biography," 253–56. Also see Crowther, *Science in Soviet Russia*, 325–36.

70. Crowther, *Science in Soviet Russia*, 175–93.

71. Crowther, *Science in Soviet Russia*, 293–306. About Levit's shtetl background and remarkable career, see James Schwartz, "Introduction," in V. V. Babkov, *The Dawn of Human Genetics* (Woodbury, NY: Cold Spring Harbor Laboratory Press, 2013), xiii–xv.

72. Viktor I. Akovlevich Frenkel, *Yakov Ilich Frenkel: His Work, Life, and Letters* (Heidelberg: Springer, 1996).

73. Victor Ya. Frenkel, "Yakov Ilich Frenkel: Sketches Toward a Civic Portrait," *Historical Studies in the Physical and Biological Sciences* 27, no. 2 (Jan. 1, 1997): 226.

74. Crowther, *Science in Soviet Russia*, 81.

75. B. I. Kochelaev, *The Beginning of Paramagnetic Resonance* (Hackensack, NJ: World Scientific Publishing, 1995).

76. Konstantin Kikoin, "Khariton, Iulii Borisovich," *YIVO Encyclopedia of Jews in Eastern Europe* (2010), www.yivoencyclopedia.org/article.aspx/Khariton_Iulii_Borisovich (accessed July 9, 2013). For a firsthand account of life and work of Khariton and others at Arzamas-16, see Veniamin Tsukerman, *Arzamas-16: Soviet Scientists in the Nuclear Age* (Nottingham, UK: Bramcote Press, 1999). For a fascinating, sustained comparison of the lives and careers of Khariton and J. Robert Oppenheimer, see David Holloway, "Parallel Lives? Oppenheimer and Khariton," in *Reappraising Oppenheimer: Centennial Studies and Reflections*, ed. Cathryn Carson and David A Hollinger (Berkeley: Office for History of Science and Technology, University of California, 2005), 115–28.

77. Slezkine, *The Jewish Century*, 221.

78. Quoted in Slezkine, *The Jewish Century*, 221. Lenin's observation explains why, in addition to the laboratory and lecture hall, Jews rose after the revolution in other literate professions and state functionary positions. The Soviet secret police (since 1934, the NKVD) illustrates this trend well: by 1937, 38% of its top leadership (42 of 111) were Jews. Seven of ten of the departments of the Main Directorate for State Security—the heart of the NKVD—were headed by Jews.

79. William Horsley Gantt, "The Soviet's Treatment of Scientists," *Current History* 31 (1930): 1151–57.

80. Peter J. Kuznick, *Beyond the Laboratory: Scientists as Political Activists in 1930s America* (Chicago: University of Chicago Press, 1987), 115–16.

81. Slezkine, *The Jewish Century*, 250.

82. Slezkine, *The Jewish Century*, 250–51.

83. David Priestland, *The Red Flag: A History of Communism* (New York: Grove Press, 2010), 282.

84. Zvi Y. Gitelman, "Soviet Antisemitism and Its Perception by Soviet Jews," in *Antisemitism in the Contemporary World*, ed. Michael Curtis (Boulder, CO: Westview Press, 1986), 189–90.

85. The official 1926 census included more than 160 ethnic nationalities in a confusing and overlapping typology. See http://demoscope.ru/weekly/ssp/ussr_nac_26.php (accessed July 30, 2012).

86. Anna Vinogradov, "Religion and Nationality: The Transformation of Jewish Identity in the Soviet Union," *Penn History Review* 18, no. 1, article 5 (Fall 2010).

87. Slezkine, *The Jewish Century*, 247.

88. Kuznick, *Beyond the Laboratory*, 117.

89. Anti-Semitism, which had never disappeared, became a barrier for Jews wishing to enter sciences during Josef Stalin's reign, a fact most graphically illustrated by the famous "doctors' plot" of 1953, in which nine doctors (six of whom were Jewish) were accused of planning to poison Soviet leaders. The charges were described in *Pravda*:

> Covering themselves with the noble and merciful calling of physicians, men of science, these fiends and killers dishonored the holy banner of science. Having taken the path of monstrous crimes, they defiled the honor of scientists. . . . The majority of the participants of the terrorist group . . . were recruited by a branch-office of American intelligence—the international Jewish bourgeois-nationalist organization called "Joint." The filthy face of this Zionist spy organization, covering up their vicious actions under the mask of kindness, is now completely revealed. ("Vicious Spies and Killers under the Mask of Academic Physicians," *Pravda*, Jan. 13, 1953, 1)

Anti-Semitism increasingly affected all the sciences after the Second World War, and this had a delayed impact, as Jews entered sciences in somewhat smaller numbers and found fewer opportunities once they did. The U.S. National Academy of Sciences had rated Soviet mathematics as "second to none" in the early 1970s, but as historian Loren Graham found, "in the late seventies some weaknesses began to show up in Soviet mathematics because of lags in computer technology and discrimination against Jewish mathematicians." Loren R. Graham, *Science in Russia and the Soviet Union: A Short History* (Cambridge: Cambridge University Press, 1993), 214.

Chapter 3. "Making a Land of Experiments"

1. Ruth Gruber, ed., *Science and the New Nations: Proceedings of the International Conference in the Advancement of New States* (New York: Basic Books, 1961).

2. Abba Eban, "Science and the New States," *Bulletin of the Atomic Scientists* 17, no. 2 (Feb. 1961): 57, 59.

3. Eban, "Science and the New States," 59–60.

4. Eban, "Science and the New States," 59. Eban did not overlook the contribution of other scholars from other countries in his account, but he did place clear emphasis on those of Israeli scientists.

5. Gruber, *Science and the New Nations*, 284–93.

6. This enthusiastic evaluation may have been particularly satisfying because Caulker had been disgruntled when the meetings started. He walked out of an early session on nuclear power, telling a reporter from the *New York Times*: "We are not quite ready for nuclear reactors. One of us is going to have to get up and say, 'Now look, everybody. Here is where we are. If you want to help us, here is where you have to start.'" Lawrence Fellows, "Israeli Scientific Meeting Aids New Nations on Modern Needs" *New York Times*, Aug. 27, 1960, 12. Tragically, Caulker died when his plane crashed upon landing in Dakar upon his return from the conference; the Weizmann Institute quickly established a scholarship in his name funding the studies of African students at the institute. "Scholarship Available at the Weizmann Graduate School in the Natural Sciences," *New Scientist* 357, no. 19 (Sept. 19, 1963): 635.

7. Eban, "Science and the New States," 59.

8. Avner Holtzman, "Lewinsky, Elhanan Leib," *YIVO Encyclopedia of Jews in Eastern Europe*, n.d., www.yivoencyclopedia.org/article.aspx/Lewinsky_Elhanan_Leib.

9. Elhanan Leib Lewinsky, *Masa' le-Erets Yisra'el bi-shenat tat (2040)* [Voyage to the Land of Israel in the year 5800 (2040)], n.d., sec. 1, http://benyehuda.org/levinsky/masa_leeretz_israel.html. Translations from this novel are mine, as are other translations from Hebrew sources unless otherwise indicated.

10. Lewisnsky, *Masa*, sec. 5.

11. Lewisnsky, *Masa*, sec. 6.

12. Lewisnsky, *Masa*, sec. 7.

13. Yaacov Shavit, "The 'Glorious Century' or the 'Cursed Century': Fin-de-Siècle Europe and the Emergence of Modern Jewish Nationalism," *Journal of Contemporary History* 26, nos. 3/4 (Sept. 1, 1991): 569. Shavit observed that "if Herzl was messianic, his was a scientific messianism. His utopia could become a reality only through scientific messianism. Fin-de-siècle Europe was a creative force which changed the world around it, and the Zionist revolution was part of this revolutionary change. His optimism, therefore, originated in a deep belief in modernity and in the new human horizons it opened up. Only against this background would the Jews be able to forge through to new horizons." Herzl provides an extreme example of an attitude that was widely shared by Zionists of almost every imaginable orientation.

14. Theodor Herzl, *The Jewish State* (1896; New York: Dover, 1988), 140.

15. Herzl, *The Jewish State*, 151.

16. Theodor Herzl and Lotta Levensohn, *Old-New Land* (*Altneuland*) (New York: Bloch, 1941), 85. For a fuller account of Herzl's modernist faith in science and its expansive place in his Zionist thought, see Jeremy Stolow, "Utopia and Geopolitics in Theodor Herzl's *Altneuland*," *Utopian Studies* 8, no. 1 (Jan. 1, 1997): 55–76.

17. Henry Pereira Mendes, *Looking Ahead: Twentieth-Century Happenings* (London: F. T. Neely, 1899), 374–77; also available online at http://archive.org/details/lookingaheadtwenoomend (accessed Nov. 11, 2012).

18. Mendes, *Looking Ahead*, 381.

19. Haim Shalom Ben-Avram, "Kommemuyot (Independence)," in *Ha-Mahar shel Etmol*, ed. Rachel Elboim-Dror, vol. 2 (Jerusalem: Yad Izhak Ben-Zvi Publications, 1993), 287.

20. Derek Jonathan Penslar, *Zionism and Technocracy: The Engineering of Jewish Settlement in Palestine, 1870–1918* (Bloomington: Indiana University Press, 1991). In this vision (as in all other things), the early Zionist thinkers were influenced by the ideological currents of Europe and North America. Edward Bellamy, whose own 1888 utopian novel *Looking Backward, 2000–1887* was a model for Mendes' *Looking Ahead*, championed technocracy, as did Thorstein Veblen in his *Engineers and the Price System* and other writings years later.

21. There were, of course, exceptions. Pessach Bar-Adon left the Hebrew University, where he had been among its first students, to apprentice himself as a Bedouin shepherd. He changed his name to Aziz Effendi. Bar-Adon (who in time devoted himself to the more scholarly pursuit of archeology) was one of a small group of immigrants who sought to recast themselves as "natives" uncorrupted by the softening temptations of the West. As one might expect, this aspiration brought reactions ranging from irritation to fury. In 1907, historian Joseph Klausner, who would later compete unsuccessfully with Chaim Weizmann to be selected as the first president of the state of Israel, disparaged this nativism from his home in Odessa: "We, Jews, who lived more than two thousand years among civilized people cannot, and should not, deteriorate again to the cultural level of semi-savage peoples." Quoted in Yosef Gorny, *Ha-Sheilah Ha-Aravit ve-ha-Beayah ha-Yehudit: Zeramim Medini'im-Ideologi'im be-Yahasom El ha-Yeshut ha-Aravit be-Ererz Yisrael* (Tel Aviv: Am Oved, 1985), 56. For a fascinating account of some of these figures, see Yael Zerubavel, "Memory, the Rebirth of the Native, and the 'Hebrew Bedouin' Identity," *Social Research* 75, no. 1 (Apr. 1, 2008): 315–52.

22. Vladimir Jabotinsky, "Doctor Herzl," http://benyehuda.org/zhabotinsky/doctor_herzl.html (accessed Nov. 25, 2012). On Jabotinsky's affection for Verne, see Michael Brown, "The New Zionism in the New World: Vladimir Jabotinsky's Relations with the United States in the Pre-Holocaust Years," *Modern Judaism* 9, no. 1 (Feb. 1, 1989): 73.

23. Max Nordau, *The Interpretation of History* (New York: Willey Book Co., 1910), 333.

24. Nordau, *The Interpretation of History*, 343.

25. Derek Jonathan Penslar, "Technical Expertise and the Construction of the Rural Yishuv, 1882–1948," *Jewish History* 14 (2000): 201. Also see Penslar's indispensable *Zionism and Technocracy*.

26. Louis Brandeis, "Palestine Has Developed Jewish Character," in *First Session of Emergency Palestine Economic Conference* (Washington, DC, 1929).

27. Louis Brandeis, "Realization Will Not Come as a Gift," in *New England Members of the Palestine Land Development League* (Boston, 1923). Already in 1915, Brandeis had written of the place of science in winning Palestine for Jews. Writing before Thanksgiving 1915, he described the travails of the "Jewish Pilgrim Fathers" who first settled in their own promised land: "The first years of these Jewish settlers resembled the first few years of the Pilgrim Fathers at Plymouth. They had to fight death and disease. Misgovernment of the country had brought malaria into it. The land appeared to be exhausted, and they knew not how to enrich and till it. Many died; and those who survived lived only to be confronted by obstacle after obstacle." Their salvation came, Brandeis continued, with the revival of Hebrew ("so that all conceptions of modern philosophy, economics, politics, science may be exprest [*sic*] in Hebrew"), the establishment of the "medical department" of the Hebrew University, and so forth. Louis Brandeis, "Democracy in Palestine," *The Independent* 84, no. 34 (Nov. 22, 1915): 311.

28. Quoted in Shulamit Reinharz, "Irma 'Rama' Lindheim: An Independent American Zionist Woman," *Nashim: A Journal of Jewish Women's Studies and Gender Issues*, no. 1 (Jan. 1, 1998): 112.

29. Irma L. Lindheim, *The Immortal Adventure* (New York: Macaulay, 1928), 89.

30. Arthur Ruppin, *The Jewish Fate and Future* (London: Macmillan, 1940), 341.

31. Still, medicine was very often the field pointed to by those who wished to demonstrate the progress that Jewish immigrants would bring to Palestine. Harry Friedenwald offers a remarkable example of this. Friedenwald was a Johns Hopkins–educated physician (and the son and grandson of prominent physicians) who became both a professor at Baltimore's College of Physicians and Surgeons and, for years, the head of the Federation of Zionists in America. During the course of his long career, Friedenwald took a long medical junket to Palestine, reporting his findings with enthusiasm upon his return. He organized at Johns Hopkins University an American exhibition of newly published Hebrew-language medical texts, attesting to the renaissance of modern Hebrew medicine. In 1896, Friedenwald had witnessed his father give a much-remarked-upon lecture on the "Jewish contribution to medicine," an experience that left him moved and changed. He himself published numerous scholarly essays on the importance of the contributions of Jews to the advancement of medicine throughout Western history and was convinced that Zionist physicians were well on the way to adding their imprint to this tradition. For an enchanting, if less than academically rigorous biography of Friedenwald, see Alexandra Lee Levin, *Vision: A Biography of*

Harry Friedenwald (Philadelphia: Jewish Publication Society of America, 1964). See, too, Friedenwald's own books in the history of Jews and medicine, all published by the Johns Hopkins Press: *Moses Maimonides, the Physician* (1935), *The Jews and Medicine* (1944), and *Jewish Luminaries in Medical History* (1946).

32. Ruppin, *The Jewish Fate and Future*, 331–32.

33. For an excellent account of Warburg's place in the promotion of Zionist technocracy (which forms the basis of the short discussion that follows), see Derek Jonathan Penslar, "Zionism, Colonialism, and Technocracy: Otto Warburg and the Commission for the Exploration of Palestine, 1903–7," *Journal of Contemporary History* 25 (1990): 143–60. For a general account of Warburg's career and associates, see Frank Leimkugel, "Botanischer Zionismus: Otto Warburg (1859–1938) und die Anfänge Institutionalisierter Naturwissenschaften in 'Erez Israel,'" *Englera*, 2005, 1–351.

34. See, for example, Otto Warburg, *Die Kautschukpflanzen und Ihre Kultur* (Berlin: Kolonial-Wirtschaftliches Komitee, 1900).

35. Penslar, "Zionism, Colonialism, and Technocracy," 152.

36. About Aaronsohn, see Shaul Katz, "On the Wings of the Brittle Rachis: Aaron Aaronsohn from the Rediscovery of Wild Wheat (Urwiezen) to His Vision for the Progress of Mankind," *Israel Journal of Plant Science* 49 (2001): S5–S17. For a less scholarly introduction to Aaronsohn's remarkable life and career from a very different perspective, see the gripping study by Ronald Florence, *Lawrence and Aaronsohn: T. E. Lawrence, Aaron Aaronsohn, and the Seeds of the Arab-Israeli Conflict* (New York: Viking, 2007).

37. David Fairchild, "An American Research Institution in Palestine: The Jewish Agricultural Experiment Station at Haifa," *Science*, n.s., 31, no. 793 (Mar. 11, 1910): 376–77. Shaul Katz, a historian of Israeli science and technology of towering reputation, long ago concluded that Aaronsohn's scientific achievements and his contribution to the practical advance of Israeli agriculture were far less than typically imagined. "His articles, which were similar to one another, teach of his broad cultural vision. They show promise, but do not contain—except in a very small way—new data or discoveries." Shaul Katz, "Aharon Aaronsohn, Reishit Ha-mada U-reishit Ha-mehkar Ha-haklai be-Eretz-Yisrael," *Kathedrah le-Toldot Eretz Yisrael* 3, no. 4 (1977 [5737]): 4. This may be true. But in helping to stitch together the practical aspirations of pioneer-farmers and the scientific aspirations of European and American researcher-agronomists, Aaronsohn played a part in fashioning the technocratic culture of Jewish settlement in Palestine in a way that may have had more influence than this or that empirical finding ever could have.

38. There are a great number of examples of efforts to bring scientific agronomy and plant biology to the Holy Land, many of which are recounted in Leimkugel, "Botanischer Zionismus."

39. Quoted in Yadin Dudai, *Ha-Mehkar ha-mada'i he'yisrael* [Scientific research in Israel] (Jerusalem, 1970), 7.

40. Quoted in Dudai, *Ha-Mehkar ha-mada'i*, 8. The text was originally writ-

ten in German and published as Martin Buber, Berthold Feiwel, and Chaim Weizmann, *Eine jüdische Hochschule* (Berlin: Jüdischer Verlag, 1902).

41. For Weizmann's wartime science and the impact that it had on the man and his careers in science and in politics, see Jehuda Reinharz, "Science in the Service of Politics: The Case of Chaim Weizmann during the First World War," *English Historical Review* 100 (1985): 572–603.

42. Hanah Arendt, "Single Track to Zion," in *The Jewish Writings*, ed. Jerome Kohn and Ron Feldman (New York: Schocken Books, 2007), 407.

43. Marshall Missner, "Why Einstein Became Famous in America," *Social Studies of Science* 15 (1985): 267–91; József Illy, ed., *Albert Meets America: How Journalists Treated Genius during Einstein's 1921 Travels*, annotated ed. (Baltimore: Johns Hopkins University Press, 2006).

44. Quoted in *Einstein on Politics: His Private Thoughts and Public Stands on Nationalism, Zionism, War, Peace, and the Bomb* (Princeton, NJ: Princeton University Press, 2007), 148.

45. Reprinted in Illy, *Albert Meets America*, 25.

46. Quoted in "Huge Throng Welcomes Visiting Zionist Leaders," *Jewish Independent*, Apr. 15, 1921, 5, reprinted in Illy, *Albert Meets America*, 85.

47. Reprinted in Illy, *Albert Meets America*, 90.

48. Reprinted in Illy, *Albert Meets America*, 97.

49. A panoramic and fascinating investigation of the establishment of the university and what followed can be found in Shaul Katz and Michael Heyd, *Toldot ha-Universitah ha-Ivrit bi-Yerushalayim: Shorashim ve-Hathalot* (Jerusalem: Magnes, 1997).

50. Universitah ha-Ivrit bi-Yerushalayim, *The Hebrew University, Jerusalem: Inauguration, April 1, 1925* (Jerusalem: Azriel Printing Works, for the Hebrew University, 1925), 5.

51. *Hebrew University Inauguration*, 27.

52. *Hebrew University Inauguration*, 24–25.

53. *Hebrew University Inauguration*, 23.

54. Louis Gershenfeld, *The Jew in Science* (Philadelphia: Jewish Publication Society of America, 1934), 206.

55. Initially, the Technicum was purely a teaching college and became a research institution only at the start of the 1950s.

56. Alex Keynan, *Science and Israel's Future: A Blueprint for Revitalizing Basic Research and Strengthening Science-Based Industry* (Jerusalem: Israel Academy of Sciences and Humanities–Jerusalem Institute for Israel Studies, 1988), 10–11. For a breezy first-person account of the institute's early years, see Joe Jaffe, *Early Days in the Weizmann Institute* (Dorset, UK: New Guild Books, 1992).

57. In the 1957–58 academic year, for instance, the Hebrew University enrolled a total of 3,998 students, of whom 812 studied science, 643 medicine, and 270 agriculture. Ritchie Calder, *Science in Israel*, vol. 7 (Jerusalem: Israel Digest, 1959), 11. In the same year, the Technion enrolled 2,300 students, all of whom

(save several dozen architecture students) concentrated in pure and applied sciences. Calder, *Science in Israel*, 18.

58. Juda Leman, *The Land of Promise*, Spielberg Jewish Film Archive (Tel Aviv: Urim Productions, 1935), pt. 49:30, www.youtube.com/watch?v=QDoD6 W2zo1s&feature=youtube_gdata_player (accessed Jan. 1, 2013).

59. Nadav Davidovich and Shifra Shvarts, "Health and Hegemony: Preventive Medicine, Immigrants, and the Israeli Melting Pot," *Israel Studies, Special Issue: Science, Technology, and Israeli Society* 9 (2004): 153.

60. Kauffmann described the "workers' village" this way:

> In the heart of the settlement, on the highest point of the hill, are found the most important cultural and economic communal buildings, crowning the settlement and at the same time outwardly embodying the principle of cooperation. Here stand the Beth Haam (People's Institute), the school, a small hospital, the central dairy, stores and sheds for agricultural produce and machinery, the Mashbir, etc., concentrating at its central point the life of the settlement. Streets or paths radiate from this point to all parts of the settlement. Between the ring formed by the farms and heart of the settlement, are placed the five-dunam homesteads of the teacher and artisans. (Richard Kauffmann, "Planning of Jewish Settlements in Palestine: A Brief Survey of Facts and Conditions," *Town Planning Review* 12, no. 2 [Nov. 1, 1926]: 110–11)

61. For an enchanting description of how similarly modernist sensibilities were reflected in the landscaping of early kibbutzim, see Elissa Rosenberg, "'An All-Day Garden': The Kibbutz as a Modernist Landscape," *Journal of Landscape Architecture* 7, no. 2 (2012): 32–39.

62. Volker M. Welter, "The 1925 Master Plan for Tel-Aviv by Patrick Geddes," *Israel Studies* 14, no. 3 (Oct. 1, 2009): 95–98.

63. Chris Renwick, "The Practice of Spencerian Science: Patrick Geddes's Biosocial Program, 1876–1889," *Isis* 100, no. 1 (Mar. 1, 2009): 47. In 1882, Darwin praised Geddes for advancing "knowledge in several branches of science." The great German evolutionary biologist August Weismann, in a letter of recommendation, wrote that "from the standpoint of general biology I can say that Mr. Geddes ranks among those of the living English [*sic*] investigators who have most deeply thought out the general biological problems which equally concern both the vegetable and the animal kingdom."

64. For a brilliant appraisal of the relationship between Bauhaus and logical positivism, with its goal of bringing scientific rigor to all realms of discourse, see Peter Galison, "Aufbau/Bauhaus: Logical Positivism and Architectural Modernism," *Critical Inquiry* 16, no. 4 (July 1, 1990): 709–52.

65. Walter Gropius, "Dessau Bauhaus: Principles of Bauhaus Production" (Dessau, Mar. 1926), quoted in Galison, "Aufbau/Bauhaus," 717.

66. Gropius, "Dessau Bauhaus," quoted in Galison, "Aufbau/Bauhaus," 717.

67. As anyone who has visited Tel Aviv knows, the impact of these International Style architects cannot be exaggerated. However, there were others who did not embrace modernism (sometimes to the ruin of their careers). About these other architects, see Alona Nitzan-Shiftan, "Contested Zionism—Alternative Modernism: Erich Mendelsohn and the Tel Aviv Chug in Mandate Palestine," *Architectural History* 39 (Jan. 1, 1996): 147–80, doi:10.2307/1568611.

68. On Bernstein, Miestechkin, and Sharon, see Nitza Metzger-Szmuk, *Des Maisons sur le Sable: Tel-Aviv, Mouvement Moderne et Esprit Bauhaus*, Edition Bilingue Français-Anglais, Eclat (Paris: Editions de l'Eclat, 2004), 128, 298, and 318–29. On Weinraub, see Richard Ingersoll, *Munio Gitai Weinraub* (Milan: Mondadori Electa, 1994). See, too, Sharon's lively and fascinating autobiographical volume, Aryeh Sharon, *Kibbutz + Bauhaus: An Architect's Way in a New Land* (Offenburg, Germany: Kramer Verlag, 1976). A good deal of information about Sharon, and enchanting photos, can be found at www.ariehsharon.org (where, for information about his Bauhaus years, see www.ariehsharon.org/BauhausDessau). A letter of reference for Weinraub signed by Mies van der Rohe is available at www.scribd.com/fullscreen/87386946?access_key=key-1b6goc54sgw36rnrd2l5 (accessed Sept. 12, 2012).

69. For more about the place of science and modernism in the imaginations of Jewish architects and planners in mandatory Palestine, see the excellent varied essays in Haim Yacobi, *Constructing a Sense of Place: Architecture and the Zionist Discourse* (Aldershot, UK: Ashgate Publishing, 2004).

70. Meron Benvenisti, *Conflicts and Contradictions* (New York: Villard Books, 1986).

71. Eliezer Smoly, *Ha-Halutzim: hokrei ha-teva shel eretz Yisrael* (Tel Aviv: 'Am 'Oved, 1972), frontispiece.

72. Israel Aharoni, *Zikhronot zu'olog Ivri* (Tel-Aviv: 'Am 'Oved, 1942/1943).

73. Aharoni, *Zikhronot zu'olog*, 1. Like so many other heroic stories that trace the development and nurturing of Palestine back only as far as the first modern Jewish settlements, this claim is inaccurate and says more about twentieth-century Zionist ideology than it does about early explorations of the land (to say nothing of the intimate knowledge of the land of those residents who predated Zionist settlement). For fascinating accounts of such earlier explorations, see Haim Goren, "Sacred, but Not Surveyed: Nineteenth-Century Surveys of Palestine," *Imago Mundi* 54 (Jan. 1, 2002): 87–110; Ruth Kark and Haim Goren, "Pioneering British Exploration and Scriptural Geography: The Syrian Society / The Palestine Association," *Geographical Journal* 177, no. 3 (Sept. 1, 2011): 264–74.

74. Aharon David Gordon, "He-Adam ve-ha-Teva," *Ben Yehuda Project*, http://benyehuda.org/gordon_ad/haadam_vehateva_02.html (accessed Sept. 12, 2012).

75. Quoted in Amos Perlmutter, "A. D. Gordon: A Transcendental Zionist," *Middle Eastern Studies* 7, no. 1 (Jan. 1, 1971): 84.

76. Quoted in Perlmutter, "A. D. Gordon," 84–85.

77. Aharon David Gordon, "Universitah Ivrit," *Ha-poel ha-tza'ir* (Mar. 31, 1925). I would like to thank Prof. Shaul Katz for introducing me to this important essay.

78. Aharon David Gordon, "Mah Mekor Ha-mevukhah Be-heshboneinu," *Ben Yehuda Project*, www.benyehuda.org/gordon_ad/ma_mekor.html (accessed Sept. 12, 2012).

79. *The Land of Promise* can be seen online at www.youtube.com/watch?v=QDoD6W2zo1s&feature=youtube_gdata_player.

80. Leman, *The Land of Promise.*

81. "The Land of Promise," *Baltimore Sun*, Feb. 4, 1936.

82. James L. Gelvin, "Zionism and the Representation of 'Jewish Palestine' at the New York World's Fair, 1939–1940," *International History Review* 22 (2000): 51–52.

83. For the relations of these fairs, especially the 1934 fair, to the emerging self-image of the Yishuv, see Sigal Davidi Kunda and Robert Oxman, "The Flight of the Camel: The Levant Fair of 1934 and the Creation of a Situated Modernism," in Haim Yacobi, *Constructing a Sense of Place*, 52–75.

84. Mordecai Naor and Batia Carmiel, *The Flying Camel: 85 Years of Exhibitions and Fairs in Tel Aviv* (Tel Aviv: Israel Trade Fairs and Convention Center & Eretz Israel Museum, 2010), 4–11 (English), 6–9 (Hebrew).

85. Naor and Carmiel, *The Flying Camel*, 7 (English).

86. Progress furniture: "Be-veit ha-haroshet le-rehitim 'Progress,'" *Davar*, Apr. 3, 1939, 4; Progress laundry soap: "Bet Franco: beit haroshet le-sabon kevisah —soda le-kevisah ve-vvkat Progress," *Davar*, July 28, 1937, 5; Progress carbonated beverages: "Hoda'ah be-inyan mekhirat nikhsei beit haroshet le-gazoz," *Davar*, Oct. 11, 1937, 5; Progress laundry services: "Progress: Beit kevisah u-tzeviah khemit modernit," *Doar ha-Yom* (Tel Aviv), Mar. 25, 1921, 1; Progress custom machining: "Niftah hadash: Beit melakhah kikhani Progress," *Davar*, Dec. 1, 1944, 5; Progress driving school: "Hoda'ah hashuvah!" *Davar*, Aug. 25, 1944, 5; English lessons at Progress Institute: "Aseh beitkha le-beit sifrekha!" *Davar*, Sept. 5, 1941, 4. The Progress furniture factory was a frequent newspaper advertiser: "Darshu he-khol batei mishar le-rehitim et ha-kisaot shel beit-ha-haroshet 'Progress,'" *Davar*, July 15, 1927, 7.

87. Naor and Carmiel, *The Flying Camel*, 42.

88. Naor and Carmiel, *The Flying Camel*, 63.

89. Naor and Carmiel, *The Flying Camel*, 74.

90. Naor and Carmiel, *The Flying Camel*, 78.

91. Quoted in Naor and Carmiel, *The Flying Camel*, 91.

92. *The Hebrew University, Jerusalem*, 108. Historian Shaul Katz has pointed out that Landau's pointed emphasis on "pure" science was part of a polemic in which he was engaged with British applied mathematician Selig Brodetsky, who eventually succeeded Magnes as president of the Hebrew University. Katz, personal communication, Aug. 23, 2011.

93. Mark Levine, "Globalization, Architecture, and Town Planning in a Colonial City: The Case of Jaffa and Tel Aviv," *Journal of World History* 18, no. 2 (June 1, 2007): 177:

In the case of Jaffa and especially Tel Aviv, the evolution of styles from garden suburb through International Style would reflect an increasing focus on defining "modern" Tel Aviv against its (apparently nonmodern) Other, although the reality is that Jaffa saw the same architectural developments as its daughter to the north. The core of the turn-of-the-twentieth-century Zionist architectural and planning imagination reflected the view of these movements at large, which increasingly leaned toward creating a "clean slate" in the (re)design of modern cities, as reflected in conferences and journals of the period. Tel Aviv's leaders shared these sentiments and specifically set out to separate their idea for a lawful, "energetic," ordered, and thus "modern" Tel Aviv from the "ugliness" and "anarchy" of Jaffa.

94. Naor and Carmiel, *The Flying Camel*, 65.

95. Arthur Hertzberg, *The Zionist Idea: A Historical Analysis and Reader* (Philadelphia: Jewish Publication Society, 1997), 134. As Shaul Katz has noted, the success of German Templer settlers (millenarian Protestants who set up in Haifa, Jerusalem, and near Jaffa in the last decades of the nineteenth century) had earlier demonstrated that the application of modern Western knowledge and techniques could improve the outcome of agricultural efforts (and building and infrastructure) to well beyond the prevailing levels in Palestine. Katz, personal communication, Aug. 23, 2011. See, too, S. Ilan Troen, *Imagining Zion: Dreams, Designs, and Realities in a Century of Jewish Settlement* (New Haven, CT: Yale University Press, 2003), 33.

96. Sander L. Gilman, *Smart Jews: The Construction of the Image of Jewish Superior Intelligence* (Lincoln: University of Nebraska Press, 1996).

97. Quoted in Noah Efron, "Trembling with Fear: How Secular Israelis View the Haredim, and Why," *Tikkun*, Sept./Oct. 1991, 89.

98. For a further discussion of this nineteenth- and early twentieth-century Jewish view of Jews, see Efron, "Trembling with Fear." Also see Rina Peled, *Ha-Adam ha-hadash shel ha-mahapekhah ha-zionit* (Jerusalem: Am Oved and the Hebrew University of Jerusalem, 2002), and the essays in Anita Shapira, *Yehudim hadashim, Yehudim yesheinim* (Tel Aviv: Am Oved, 1997).

99. Francis Galton, "Eugenics: Its Definition, Scope, and Aims," *American Journal of Sociology* 10, no. 1 (July 1, 1904): 1.

100. Raphael Falk, "Zionism, Race, and Eugenics," in *Jewish Tradition and the Challenge of Darwinism*, ed. Geoffrey Cantor and Marc Swetlitz (Chicago: University of Chicago Press, 2006), 151. See, too, the fascinating book-length treatment in Raphael Falk, *Zionut ve-ha-biologiah shel ha-yehudim* (Tel Aviv: Resling, 2006).

101. Quoted in Calder, *Science in Israel*, 7:6.

102. Quoted in Shalom Reichman, Yossi Katz, and Yair Paz, "The Absorptive Capacity of Palestine, 1882–1948," *Middle Eastern Studies* 33, no. 2 (Apr. 1, 1997): 341. The translation is theirs. The original can be found online at www .benyehuda.org/ginzberg/Gnz002.html (accessed Sept. 15, 2012).

103. For an excellent account, see Ilan Troen, "Calculating the 'Economic Absorptive Capacity' of Palestine: A Study of the Political Uses of Scientific Research," *Contemporary Jewry* 10, no. 2 (Fall 1989): 19–38. See, too, Reichman, Katz, and Paz, "The Absorptive Capacity of Palestine, 1882–1948." A fascinating survey of Taylorism and its importance to Jews wishing to settle Palestine is given in "Mada ha-avodah ve-hisakhon ba-zman," *Davar*, June 3, 1925, 2.

104. Quoted in Efraim Katzir, "Reishito Shel ha-Mo'etzah ha-Mada'it—Ben Guryon ve-ha-Hemed," in *David Ben-Guryon Ve-hitpathut Ha-mada` be-Yi´sra'el: Yom `iyun Bi-melot Me'ah Shanah Le-huledet David Ben-Guryon* [David Ben Gurion and the development of science in Israel: A symposium commemorating the hundredth anniversary of Ben Gurion's birth, held 23 April, 1987] (Jerusalem: Publications of the Israel Academy of Sciences and Humanities, 1989), 28.

105. Avner Cohen, "Before the Beginning: The Early History of Israel's Nuclear Project (1948–1954)," *Israel Studies* 3, no. 1 (Apr. 1, 1998): 112–39; Avner Cohen, *Israel and the Bomb* (New York: Columbia University Press, 1998).

106. Ami Maoz, "Ha-havnayah ha-hevratit shel proyekt yibush ha-hula, 1948–1952" [The social construction of the Hula Drainage Project, 1948–1952] (M.A. thesis, Bar Ilan University, 2010).

107. Benita Tall, "Land of Milk, Honey and Science," *Science News-Letter* 75, no. 16 (Apr. 18, 1959): 250–51.

108. Golda Meir, *My Life* (London: Weidenfeld and Nicolson, 1975), 263–90; Ethan A. Nadelmann, "Israel and Black Africa: A Rapprochement?," *Journal of Modern African Studies* 19, no. 2 (June 1, 1981): 184–88.

109. Leopold Laufer, "Israel and the Third World," *Political Science Quarterly* 87 (1972): 619.

110. Michael Brecher, "Israel and 'Afro-Asia,'" *International Journal* 16 (1961): 107–37; Bernard Reich, "Israel's Policy in Africa," *Middle East Journal* 18 (1964): 14–26; Fouad Ajami and Martin H. Sours, "Israel and Sub-Saharan Africa: A Study of Interaction," *African Studies Review* 13 (1970): 405–13.

111. Quoted in A. B. Zahlan, "The Science and Technology Gap in the Arab-Israeli Conflict," *Journal of Palestine Studies* 1 (1972): 3.

112. The literature on the roles of Western science and technology in colonial enterprises is expansive, unruly, and fascinating. See, for example, Laurelyn Whitt, *Science, Colonialism, and Indigenous Peoples: The Cultural Politics of Law and Knowledge* (Cambridge: Cambridge University Press, 2009); Suman Seth, "Putting Knowledge in Its Place: Science, Colonialism, and the Postcolonial," *Postcolonial Studies* 12, no. 4 (2009): 373–88; Daniel R. Headrick, *Power over Peoples: Technology, Environments, and Western Imperialism, 1400 to the Pres-*

ent (Princeton, NJ: Princeton University Press, 2012); Joseph Morgan Hodge, *Triumph of the Expert: Agrarian Doctrines of Development and the Legacies of British Colonialism* (Athens: Ohio University Press, 2007); Sandra Harding, *The Postcolonial Science and Technology Studies Reader* (Durham, NC: Duke University Press, 2011).

113. Dan Senor, quoted in Dwyer Gunn, "How Did Israel Become 'Start-Up Nation'?," *Freakonomics: The Hidden Side of Everything* website, Dec. 4, 2009, www.freakonomics.com/2009/12/04/how-did-israel-become-start-up-nation (accessed Jan. 1, 2013). The book is Dan Senor and Saul Singer, *Start-Up Nation: The Story of Israel's Economic Miracle* (New York: Twelve, 2009).

114. Quoted in Yuri Yalon, "Mokirim todah la-kippah," *Yisrael ha-Yom* (Tel Aviv), Sept. 3, 2012, 12.

115. Quoted in Calder, *Science in Israel*, 7:5.

Conclusion. When All Worlds Were New Worlds

1. Yuri Slezkine, *The Jewish Century* (Princeton, NJ: Princeton University Press, 2004), 207. Still, there is anachronism in Slezkine's geography. In 1926, David Bergelson published a much-discussed essay called "Dray Tsentern" [Three centers] which considers the natures and virtues of what Bergelson took to be, undeniably, the three hubs of contemporary Jewish life: the United States, Poland, and the Soviet Union. David Bergelson, *Descent* (New York; Modern Language Association, 1999), xxi–xxii.

2. "Schiff Averts Panic: Rostrum Breaks Down; Rush of Jews to Greet Dr. Lewin Overloads Platform at Durland's," *New York Tribune*, Dec. 10, 1906.

3. "Throng of Thousands Greets Rabbi Lewin," *New York Times*, Dec. 10, 1906.

4. "Throngs of Thousands."

5. "Schiff Averts Panic."

6. "Schiff Averts Panic."

7. "Rousing Farewell Given to Dr. Lewin: Big Gathering of Zionists Hears His Speech for a Jewish Nation," *New York Times*, Feb. 11, 1907.

8. "Rousing Farewell." Also see "Mass Meeting in Drive Today: Jewish Community to Gather at Trinity Auditorium; Dr. Schmarya Levin Will Be Principal Speaker; Regarded as Leader in Move for Zionism," *Los Angeles Times*, June 14, 1928.

9. So popular was Weizmann in the United States that when he came with Einstein in 1921 as part of the Zionist delegation charged with raising funds for Palestine, he was met by raucous overflow crowds of tens of thousands wherever he went. One scholar has gone so far as to suggest that Einstein came to be seen as a popular hero in America because the local non-Jewish press mistook the unprecedented enthusiasm of American Jews for Weizmann as a tribute to Einstein. See Marshal Missner, "Why Einstein Became Famous in America," *Social Studies of Science* 15 (1985): 267–91.

10. This traffic also took forms that were more homely, though no less important. For instance, Abraham Cahan told an interviewer in 1911 that "in these modern days, there are hundreds of Russian towns, the bulk of whose mail is from America," referring to the correspondence from Russian Jews who moved to cities like New York to relatives who moved to cities like Saint Petersburg. Ernest Poole, "Abraham Cahan: Socialist—Journalist—Friend of the Ghetto," *Outlook*, Oct. 28, 1911, 470.

11. See, for example, Angelo S. Rappoport, "The Russian Douma and the Emancipation of the Jews," ed. William Leonard Courtney, *Fortnightly Review* 89, no. 532 (Apr. 1911): 648–61.

12. "Russian Jews' Bund Greets Jews Here: Tells of Its Aims Now That It Is Free to Work in the Open; Plans National Congress; Says First Problem Is Prompt Organization of the Democratic Self-Ruling Institutions," *New York Times*, Apr. 21, 1917.

13. Henry Pereira Mendes, *Looking Ahead: Twentieth-Century Happenings* (London: F. T. Neely, 1899), 381; available online at http://archive.org/details/lookingaheadtwenoomend.

14. Bernard G. Richards, "Zionism and Socialism," *Arena*, Mar. 1903, 276.

15. Poole, "Abraham Cahan," 472.

16. Poole, "Abraham Cahan," 472

17. The same could have been said for most of the grandparents themselves, as the nineteenth century was for a great many European Jews a time of upheaval and durable change as well. This fact, of course, only amplified the degree to which early twentieth-century Jews, in a great many places in the West, faced a future that they knew could not resemble their past.

18. Hyman Levy, *The Universe of Science* (New York: Century Publishing, 1933), 189.

19. Robert King Merton, "Science and the Social Order," *Philosophy of Science* 5, no. 3 (July 1938): 326–27.

20. Franz Boas, "Race Prejudice from the Scientist's Angle," *Forum and Century*, Aug. 1937, 94.

21. J. Robert Oppenheimer, "Encouragement of Science," *Science News-Letter* 57, no. 11 (Mar. 18, 1950): 170.

22. Slezkine, *The Jewish Century*, 250–51.

23. Universitah ha-Ivrit bi-Yerushalayim, *The Hebrew University, Jerusalem: Inauguration, April 1, 1925* (Jerusalem: Azriel Printing Works, for the Hebrew University, 1925), 24–25.

24. David Ben Gurion, "Israel's Fourteen Years," *New York Herald Tribune*, May 15, 1962, 1, 12.

25. Yuri Slezkine has offered an explanation of a different sort for the fact that Ashkenazi Jews throughout the West displayed like enthusiasm for twentieth-century science. As he sees it, twentieth-century science, like so much else about

the epoch, both demanded and fostered precisely those traits that generations of history had impressed upon Jews. "The Modern Age is the Jewish Age," Slezkine writes, "and the twentieth century, in particular, is the Jewish Century."

> Modernization is about everyone becoming urban, mobile, literate, articulate, intellectually intricate, physically fastidious, and occupationally flexible. It is about learning how to cultivate people and symbols, not fields or herds. It is about pursuing wealth for the sake of learning, learning for the sake of wealth, and both wealth and learning for their own sake. It is about transforming peasants and princes into merchants and priests, replacing inherited privilege with acquired prestige, and dismantling social estates for the benefit of individuals, nuclear families, and book-reading tribes (nations). Modernization, in other words, is about everyone becoming Jewish. Some peasants and princes have done better than others, but no one is better at being Jewish than the Jews themselves. In the age of capital, they are the most creative entrepreneurs; in the age of alienation, they are the most experienced exiles; and in the age of expertise, they are the most proficient professionals. Some of the oldest Jewish specialties—commerce, law, medicine, textual interpretation, and cultural mediation—have become the most fundamental (and the most Jewish) of all modern pursuits. It is by being exemplary ancients that the Jews have become model moderns. (Slezkine, *The Jewish Century*, 1)

Slezkine is right in observing links between new ways science was pursued in the twentieth century and new ways in which business, government, and practically everything else was pursued at the same time. And he is right that traits that many Jews in many places had long displayed—literacy, mobility, a tendency to congregate in cities, a facility with *Luftgeschäft*, and so on—were traits that twentieth-century science demanded of its practitioners. There is brilliance in Slezkine's explanation, and valuable insight.

At the same time, at least as far as modern science goes, Slezkine's explanation accounts for less than it perhaps seems to at first. The enthusiasm that many Jews displayed toward twentieth-century science did not always fit snugly with traits and predilections they already had. In many cases, as I have tried to demonstrate above, their enthusiasm was because twentieth-century science promised to perform a dual reformation—rendering Jews more fit for the societies in which they found themselves, and rendering these same societies more fit for Jews. Slezkine's account may pay too little attention to Jews' radically aspirational nature of the embrace of science (and perhaps other archetypally modern institutions like film and literature, or media). It was not simply that in the twentieth century, societies throughout the West *converged* on the complex of what Slezkine called mercurian traits that Jews long ago made their own. It was that in the twentieth century, a great many Ashkenazi Jews, in a great many different places, devoted themselves with unparalleled passion and purpose to remaking both themselves and the places they lived and loved in a new image. That their complicated histories provided many of them with cultural resources to effect this refashioning with a good deal of success may account to some degree for the remarkable achievements of Jews in the sciences. (For my breathlessly rhapsodic review of Slezkine's book, see Noah Efron, "Some of My Best Friends Are Mercurians," *Jerusalem Report*, Jan. 24, 2005.)

26. Quoted in "Nazi's Conception of Science Scored: 1,284 American Scientists Sign Manifesto Rallying Savants to Defend Democracy," *New York Times*, Dec. 11, 1938, 50. Boas' quotation is taken from a resolution passed by the American Association for the Advancement of Science in December 1938, condemning Nazi discrimination against non-Aryan scientists.

27. Jean Paul Sartre, *Anti-Semite and Jew* (1948; New York: Schocken, 1995), 110–11.

28. Gustav Ichheiser, "Antisemite and Jew," *American Journal of Sociology* 55, no. 1 (July 1, 1949): 110–12.

29. Octavio Paz, "Sartre in Our Time," *Dissent* 27, no. 4 (Fall 1980): 429.

30. For the disproportionate representation of Jews in the Soviet police (which was very similar in extent to their disproportionate representation in sciences), see Slezkine, *The Jewish Century*, 254–55.

31. The four letters Einstein sent to Roosevelt are available online at http://hypertextbook.com/eworld/einstein.shtml (accessed Oct. 10, 2012).

32. Yaron Ezrahi, "Necessary Fictions: The Decline of Science in the Democratic Imagination," Science and Democracy Lecture, Harvard University, April 9, 2007, www.youtube.com/watch?v=oxz2rKwnDaU&feature=youtube_gdata_player. For an outstanding and surprising description of the ways science is deployed in democracies of our day, see also Yaron Ezrahi, *Imagined Democracies: Necessary Political Fictions* (Cambridge: Cambridge University Press, 2012).

33. Andrew Jewett, "Science and the Promise of Democracy in America," *Daedalus* 132 (2003): 70. More recently, Jewett has put the matter this way:

> By the mid-1960s, a new generation of critics viewed with alarm a series
> of interrelated phenomena: the increasing interpenetration of science,
> industry, foundations, and the state, aimed at both military superiority
> and consumption-driven economic growth; the vigorous claims of value-
> neutrality and political detachment by researchers who worked to meet
> the instrumental needs of the governing complex; and the image—and
> to some extent the reality—of democracy as a process in which experts
> and bureaucrats simply handed down decisions and material benefits to
> passive, naïvely trusting citizens. Science lay at the heart of the "system"
> targeted by 1960s critics, who grasped all too well the ideological and
> practical centrality of science to postwar American governance. The
> epistemological claims and political ties of the postwar scientific establish-

ment led many on the left to conclude that a stance of self-professed detachment was hypocritical and pernicious in its social and political effects.

The notion that science was both an archetype and an agent of democratic political culture had, by this time, become controversial, if not outright implausible. Andrew Jewett, *Science, Democracy, and the American University: From the Civil War to the Cold War* (Cambridge: Cambridge University Press, 2012), 365.

34. Noah Efron, "American Jews and Intelligent Design," Reilly Center Reports, University of Notre Dame, vol. 1 (2013), http://reilly.nd.edu/assets/65756/rcrefron.pdf (accessed July 8, 2013).

35. "A Nobel for Negligence?," Haaretz.com, Oct. 13, 2009, www.haaretz.com/print-edition/opinion/a-nobel-for-negligence-1.6218 (accessed Oct. 10, 2012).

Hess, Moses, 87
Higham, John, 17
Higher Learning in America, The
  (Veblen), 1
high-tech, Israeli involvement in, 92–93
Hirsch, Francine, 41
Hollinger, David, 6, 7
homeland, service to, 10
Howells, William Dean, 18
How the Other Half Lives (Riis), 18
Hula Swamp, 90
human character, plasticity of, 40
humanity, service to, 10

IDF. *See* Israel Defense Forces
immigrants, assimilation of, 21
Imperial Academy of Sciences, 51
Institut für Kolonialwirtschaft (Institute
  for Colonial Science), 73
Institute for Advanced Study (Princeton,
  NJ), 25, 26
Institute for Chemical Physics, 58
Institute for Fibers and Forest Products
  (Israel), 66
Institute for Physical Problems, 58
"Intellectual Pre-eminence of Jews in
  Modern Europe, The" (Veblen), 2
intelligent design, 33, 103–4, 118–19n90
International Conference on the Role of
  Science in the Advancement of New
  States, 64–66
*International Jew, The: The World's
  Foremost Problem*, 20
international Jews, stereotype of, 17
International Style, 80–81
Interpretation of History (Nordau),
  71–72
Ioffe, Abram Fyodorovich, 54–58, 62
Iron Dome, 93
irony, 2, 10
Ish-Horowitz, Samuel Joseph, 88
Israel: establishment of, 90; offering
  scientific assistance to developing
  nations, 66; relations with the Third
  World, 91; science and technology
  essential to, 93; science's decline
  in, 104; science's role in society
  of, 66–67; scientific planning and
  development in, 65–66; Soviet Jews'

immigrating to, 63; as start-up nation,
  92; technological prowess of, 90–91;
  unique role of, for bridging the developed and developing world, 65–66;
  as a Western country, 92. *See also*
  Palestine
Israel Atomic Energy Commission, 66
Israel Defense Forces, 90

Jabotinsky, Ze'ev (Vladimir), 71, 86
Jacobs, Joseph, 5
Jewett, Andrew, 29, 103, 117–18n70,
  140–41n33
Jewish Academy of Arts and Sciences, 16
Jewish Immigrants' Information Bureau,
  21
"Jewish Institute of Higher Education,
  A" (Weizmann et al.), 75
Jewish National Fund, 86
Jewish Palestine Pavilion (NY World's
  Fair, 1939), 84
Jewish Problem (Jewish Question), 8–9
Jews: analytical style of, 4; assimilation
  of, 10–11; bookishness of, overes-
  timated, 95–96; bringing modern
  civilization to the Levant, 87; coming
  to science in the 20th century, 5;
  commonality of experience among, 8;
  communities of, linked, 96–97; declining success in science, 6; embracing
  revolution, 46–47; fertility of scholars,
  3; genetically predisposed to success
  in science, 3, 5; intellectual achievement of, sociobiological theory of,
  3–4; measures of, in science, 15;
  migration of, 7, 16–17, 45; nationalist awakening of, 72; new worlds for,
  97, 99, 102; official status of, after the
  Russian Revolution, 51–52; place of,
  in Western society, 8, 10–11; poverty
  of, after immigrating, 18; reform of,
  9; represented disproportionately
  in Western science, 2–3, 6; Russian,
  before the Revolution, 42–47; scholarly tradition of, 4, 5; shared experiences of, 99–100; talents of, compared
  with other Westerners, 5. *See also*
  American Jews; Soviet Jews
Jews in American History (Golden), 30

science (*cont.*)
  role in Western societies, 103; serving
    political Zionism's ideological agenda,
    89; as tool for entering non-Jewish
    society, 23; traits required for, 2; unre-
    lated to the individual, 100
Science and the Public Mind (Gruen-
  berg), 34
scientific democrats, 29–30
scientific messianism, 68
scientific modernism, 80, 81
scientists: drawn to socialism, 57; as folk
  heroes for US Jews, 14–15; status of,
  103
Scopes trial, 12–14
Sectarian Invasion of Our Public Schools,
  The (Newman), 31
Semenov, Nikolai, 58
Shapira, Zvi Herman, 74–75
Sharett, Moshe, 91
Sharon, Arieh, 81
Sieff (Daniel) Research Institute, 79
Silver, Abba Hillel, 76
skepticism, 2
Slezkine, Yuri, 7–8, 60, 61, 62, 94,
  138–40n25
Slosson, Edwin Emery, 77
Smolenskin, Peretz, 42
Snow, Charles Percy, 2–3
socialism, appeal of, 46–47
society, refashioning, through science,
  9–10, 14
Society for the Study of Social Biology
  and Psycho-Physics of the Jews, 39
socio-historical psychology, 40
Sokolov, Nahum, 86
Soskin, Selig, 74
Soviet Academy of Sciences, 58
Soviet development psychology, 40
Soviet Jews: education of, 53; evolving
  with Soviet science, 41–47, 53–54;
  immigrating to Israel, 63; importance
  of, to the revolution, 60; Jewishness
  of, 62; nationhood of, 62; not associated with tsarist past, 60; occupations
  of, 53; spurred toward the sciences,
  61–63
"Soviet's Treatment of Scientists, The"
  (Gantt), 60–61

Soviet Union: civilizing the masses in,
  40–41; collapse of, 104; corruption
  and oppression in, 63; as destination
  for Jews, 7; Great Purge in, 61; intellectuals in, treated with suspicion, 57;
  internal passports in, identifying Jewish nationality, 62; Jews as subjects of
  study in, 39–40; Jews overrepresented
  in science in, 8; Jews' changed status
  in, 51–52; needing scientific infrastructure, 59–60; science in, entwined
  with Soviet Jewry, 41–42, 53–54;
  scientific outlook of, 8, 62–63. *See
  also* Russia
Sprat, Thomas, 37
Staatliches Bauhaus, 80–81
State Publishing House, 58
State Roentgenological and Radiological
  Institute (GRRI; USSR), 56–57, 58
Steiner, George, 4
Stolypin's Coup, 49
Sylvester, James Joseph, 113n
Synagogue Sisterhoods, 16
Szold, Henrietta, 72

Talmud study, 4–5
Tamm, Igor Yevgenyevich, 54, 58
Technicum (Technion), 79, 97
Tel Aviv: architecture in, 135n93; fairs
  and exhibitions in, 84–86; master plan
  for, 80–81
Teller, Edward, 54
Theory and Practice (Karp), 16
Theory of the Leisure Class, The
  (Veblen), 1
Theremin, Leon (Lev Sergeyevich Ter-
  men), 56
Timiryazev, Kliment Arkadievich, 47,
  48, 49
Tolstoi, I. I., 49
Triedel, Joseph, 74
Turing (A. M.) Award, 15

United Orthodox Jewish Congregations
  of Cleveland, 31
United States: anti-Semitism in, 17–18,
  20, 21–22; as destination for Jews, 7,
  16–17; Jews overrepresented in science in, 8; making it in, 21–22; public

schools' role in, 30–32; scientific
values in, 8
U Nu, 91
Ussishkin, Menachem Mendel, 25, 76

Veblen, Thorstein, 1–2, 5, 6, 10
Vernadsky, Vladimir I., 48, 50, 51
Vygotsky, Lev Semionovich, 40

Warburg, Otto, 73–74
War of the Worldviews (Ham), xi
Weinraub, Munio, 81
Weizmann, Chaim, 25, 44, 75–78, 79,
  86, 88, 89, 90, 92, 93, 97, 100
Weizmann Institute of Science, 79
Western society, Jews' place in, 2, 8,
  10–11
Weyl, Nathaniel, 3
Who's Who in American Jewry (1938), 15
Wiener, Norbert, 3
Williams, William Carlos, 25–26
Wise, Stephen, 32
Workers' Opposition, 57
World's Fair (New York, 1939), 84

World War I, 2, 50
World Zionist Organization, 75; Commission for the Exploration of Palestine, 74; Palestine Bureau, 72–73
WPA Art Project, 16

Yiddish: culture of, in the Soviet Union,
  62; modernizing culture of, 46; science books in, 16

Zeta Beta Tau, 36
Zionism, 2, 87; associated with scientific
  achievement, 91; associated with the
  West, 89; consensus on, limited, 71;
  displacing the locals, 87; eugenics
  and, 89; facets of, 70–71; immigration waves of, 70; liberal technocracy
  of, 71–72; progressive nature of, 86;
  prose of, yearning for normalcy, 88;
  social engineering of, 70; support
  for, and scientific success, 76–77;
  technological and scientific planning
  as elements of, 66–72
Zionist Congresses, 71, 75